Crypto-Politics

This book examines current debates about the politics of technology and the future of democratic practices in the digital era.

The volume centres on the debates on digital encryption in Germany and the USA during the aftermath of Edward Snowden's leaks, which revolved around the value of privacy and the legitimacy of surveillance practices. Using a discourse analysis of mass media and specialist debates, it shows how these are closely interlinked with technological controversies and how, as a result, contestation emerges not within one public sphere but within multiple expert circles. The book develops the notion of 'publicness' in order to grasp the political significance of these controversies, thereby making an innovative contribution to Critical Security Studies by introducing digital encryption as an important site for understanding the broader debates on cyber security and surveillance.

This book will be of much interest to students of Critical Security Studies, Science and Technology Studies, and International Relations.

Linda Monsees is a Postdoctoral Fellow at the Goethe University Frankfurt, Germany.

Routledge New Security Studies

The aim of this book series is to gather state-of-the-art theoretical reflection and empirical research into a core set of volumes that respond vigorously and dynamically to new challenges to security studies scholarship. This is a continuation of the PRIO New Security Studies series.

Series Editors: J. Peter Burgess, École Normale Superieur (ENS), Paris

Surveillance, Privacy and Security
Citizens' perspectives
Edited by Michael Friedewald, J. Peter Burgess, Johann Čas, Rocco Bellanova and Walter Peissl

Socially Responsible Innovation in Security
Critical Reflections
Edited by J. Peter Burgess, Genserik Reniers, Koen Ponnet, Wim Hardyns and Wim Smit

Visual Security Studies
Sights and Spectacles of Insecurity and War
Edited by Juha A. Vuori and Rune Saugmann Andersen

Privacy and Identity in a Networked Society
Refining Privacy Impact Assessment
Stefan Strauß

Energy Security Logics in Europe
Threat, Risk or Emancipation?
Izabela Surwillo

Crypto-Politics
Encryption and Democratic Practices in the Digital Era
Linda Monsees

Negotiating Intractable Conflicts
Readiness Theory Revisited
Amira Schiff

For more information about this series, please visit: https://www.routledge.com/Routledge-New-Security-Studies/book-series/RNSS

Crypto-Politics

Encryption and Democratic Practices in the Digital Era

Linda Monsees

Routledge
Taylor & Francis Group

LONDON AND NEW YORK

First published 2020 by Routledge

2 Park Square, Milton Park, Abingdon, Oxon, OX14 4RN

605 Third Avenue, New York, NY 10017

Routledge is an imprint of the Taylor & Francis Group, an informa business

First issued in paperback 2020

British Library Cataloguing in Publication Data
A catalogue record for this book is available from the British Library

Library of Congress Cataloging-in-Publication Data
Names: Monsees, Linda, author.
Title: Crypto-politics : encryption and democratic practices in the digital
    era / Linda Monsees.
Description: Abingdon, Oxon ; New York, NY : Routledge, 2019. |
    Series: Routledge new security studies | Includes bibliographical
    references and index.
Identifiers: LCCN 2019009960 (print) | LCCN 2019013512 (ebook) | ISBN
    9780429852688 (Web PDF) | ISBN 9780429852664 (Mobi) |
    ISBN 9780429852671 (ePub) | ISBN 9781138314788 (hardback) |
    ISBN 9780429456756 (ebk.)
Subjects: LCSH: Data encryption (Computer science)--Political
    aspects--United States. | Data encryption (Computer science)--Political
    aspects--Germany. | Computer security--Political
    aspects–United States. | Computer security--Political aspects--Germany.
    | Privacy, Right of--United States. | Privacy, Right of--Germany.
Classification: LCC QA76.9.D335 (ebook) | LCC QA76.9.D335 M66 2019
    (print) | DDC 005.8/24--dc23
LC record available at https://lccn.loc.gov/2019009960

ISBN: 978-1-138-31478-8 (hbk)
ISBN: 978-0-367-78518-5 (pbk)

Typeset in Times New Roman
by Taylor & Francis Books

Contents

	List of tables	vi
	Acknowledgements	vii
1	Crypto-politics	1
2	Researching the politics of technology	18
3	On publics in the technological society	38
4	Digital encryption	58
5	Encryption and national security	80
6	Contentious technology	111
7	The politics of publicness	132
	Appendix	144

Tables

5.1 The state in control   87
5.2 The state out of control   88

Acknowledgements

This book was finalised while being a Post-Doctoral Researcher at the Cluster of Excellence 'Normative Orders' at the Goethe University Frankfurt, which provided an exceptional intellectual environment for my work. I am grateful for the feedback provided by Andreas Baur, Javier Burdman, Malcolm Campbell-Verduyn, Eva Johais, Matthias Leese, Phillip Olbrich and the 'Normative Orders' IR colloquium. Jef Huysmans offered guidance throughout the whole process and Peter Burgess's generous feedback helped to finalise the manuscript. All mistakes, inconsistencies and gaps are solely my responsibility.

1 Crypto-politics

Encryption is a core technology for internet security. It secures data against unwanted access by third parties, for example by keeping email private and preventing digital espionage. This book examines digital encryption as a security technology and analyses the controversies revolving around it. Digital encryption is a fundamental political technology; a site where struggles about privacy, freedom and democracy are fought. In a pamphlet published in 1992, author and tech expert Timothy C. May described encryption as a contentious but also visionary technology:

> A specter is haunting the modern world, the specter of crypto anarchy. Computer technology is on the verge of providing the ability for individuals and groups to communicate and interact with each other in a totally anonymous manner [...] The state will of course try to slow or halt the spread of this technology, citing national security concerns [...] Various criminal and foreign elements will be active users of CryptoNet. But this will not halt the spread of crypto anarchy.
>
> (May, 1992)

May envisions encryption as a technology endowed with the potential of radically changing existing forms of political ordering. What he calls 'crypto anarchy' is a world in which, through the power of encryption, the power of the state is challenged in a fundamental way. Today, encryption is a technology that is indeed used on a daily basis. But the transformative impact of encryption that May envisioned is far from self-evident. Developing sophisticated encryption systems and, in turn, software to break or circumvent them is big business. From this perspective, digital encryption is less a technology that transforms power relations than a means for stabilising the power of the state. So how is it that encryption, which was once seen as an utopian tool for an alternative world order, is now part of a rapidly growing security industry? Can we still see glimpses of May's visionary ideas today, and do encryption controversies enact an alternative kind of politics? *Crypto-Politics* sets out to answer these questions.

The first question is answered through an empirical analysis of controversies (Barry, 2012; Callon *et al.*, 2009; Schouten, 2014). I investigate how debates about encryption technology are entangled with controversies around security, surveillance and privacy. In Chapters 4, 5 and 6, I describe various instances of how encryption becomes contested and scrutinise multiple controversies concerned with security, technological features and the role of companies in providing encryption. What emerges is that encryption is still envisioned as a potent tool against state; a means to protect against surveillance by the state. A countervailing narrative promoted by state actors depicts it as a tool that hampers law enforcement. From this perspective, strong encryption causes insecurity by making communication among criminals inaccessible. The answer to the second question is provided by a theoretical investigation of how we can observe modes of 'publicness'. In particular, I discuss how these modes of publicness challenge or reaffirm ideas about what security is, and how they acquire political significance. In Chapter 3, I introduce the notion of *publicness*, which provides the analytical leverage to understand controversies as potential forms of 'radical'[1] contestation. In Chapter 7, I rely on this theoretical building block and the empirical analysis to reflect further on the ways in which encryption controversies enact politics. Ultimately, I show how encryption controversies challenge established ideas of the state as the prime provider of security, but also reaffirm political ideas such as the distinction between citizens and non-citizens.

The explicit focus of this book is on digital encryption and security. The issue of digital encryption has so far received little attention in International Relations (although see the discussions in: Dunn Cavelty, 2007; Herrera, 2002; Saco, 2002). However, encryption is a crucial aspect of debates about cybersecurity[2] more generally, and surveillance and privacy specifically. The realm of security is not only of particular empirical interest. Analysing security is also crucial because claims to security rely on specific imaginaries of the political. For example, security claims always enact specific claims about the role of the state (Huysmans, 2011). The state is often considered to be the prime provider of security, but in other contexts it is also considered to be a threat (Walker, 1997: 63). This tension becomes apparent in encryption controversies, where this opposition is a dominant theme. For this reason, this book looks at encryption technology as a security technology from a state perspective, but also investigates how activists contest state policies and how users are entangled in the politics of encryption. This study is based on a qualitative analysis of texts produced by activists, politicians, experts and state agencies. The focus is on the debates that started in the aftermath of the revelations by Edward Snowden. These texts were analysed in order to understand the controversies that revolve around encryption. In Chapter 2, I describe my method of analysis and the material on which that analysis is based in more detail. *Crypto-Politics* thus expands our empirical knowledge of encryption as a crucial security technology and combines it with a theoretical reflection about how the contestation of security technology works inside and outside established political categories.

The rest of this introductory chapter will introduce the main themes of the book, starting with a brief description of encryption as a security technology, pointing out its relevance for debates on surveillance and privacy. I will then introduce some conceptual arguments, on which I will draw for the study of security technology throughout the book. Next, I introduce the particularly important concept of 'publicness', which can help us to grasp the political dimension of security controversies, before concluding with a brief chapter outline.

Debating encryption

Controlling access to and implementation of digital encryption means controlling who has access to what kind of information. This is so because encrypting information means rendering it unintelligible to third parties who lack the necessary key to decrypt the information. The capabilities of encryption increased in importance with the advent of the internet. Today, encryption is used whenever we write an email, pay by credit card or even use a remote control. Encryption ciphers communication and prevents eavesdropping, but it can also function as a digital signature, thus verifying the origin of a message. The importance of encryption will only increase in the future with the expansion of the internet of things, in which many mundane devices become networked. Securing self-driving cars or networked health devices requires extremely good encryption.

The political debates about encryption can be traced back to the early days of the internet. These issues will be described in more detail in Chapters 4 and 5; here, I will merely introduce the main conflict lines. During the 1990s, encryption became the battleground for a fundamental conflict between 'hackers' and civil society organisations, on one side, and the US government and its law enforcement agencies, on the other (for an overview, see: Diffie and Landau, 1998; Kahn, 1997). With the development of digital communication, state efforts to control this communication and have access to it increased. This led to two important questions. Who should have access to the strongest kind of encryption? And who should decide on the design and implementation of encryption? Activists advocated for the free use of encryption in order to promote security and privacy, while law enforcement agencies wanted to weaken available encryption in order to maintain their ability to wiretap all communication (Hellegren, 2017; Levy, 1994). In the end, the state lost, in part because business had a vested interest in strong encryption (Herrera, 2002: 113–117; Saco, 2002: 178).

The conflict over encryption is not only one between the state and civil society. Commerce adds another layer of interest to the debate, because strong encryption is considered to be crucial for all financial transactions and to prevent industrial espionage.[3] In addition, if encryption is forbidden in only one country, that country will fear losing out in the global marketplace. Thus, economic interests aligned with the interests of activists in their fight for the spread of strong encryption. Nevertheless, companies such as AT&T also cooperated with the US government to produce end devices that implemented weak encryption. Today, Google has to prevent end-to-end encryption to sustain its business model, which is based on targeted advertising and Dropbox, for example, allows encryption, but this massively increases the cost of data storage (see DuPont, 2016). Although the power of economic actors is crucial for the issue of encryption, their interests differ depending on the impact of encryption on their business models. Hence, the role of information and communication technology (ICT) companies is often ambiguous.

Struggles over encryption are embedded in broader societal questions about control over networked technology and private data. For example, state-led surveillance practices became a major public issue after Snowden's revelations, which rekindled the debate about encryption (Gürses *et al.*, 2016; Hintz and Dencik, 2016; Kuehn, 2018). Encryption can hamper some surveillance practices and plays a pivotal role in digital anonymisation technologies. Implementing strong encryption is thus one way to complicate surveillance efforts by both the state and business. Civil rights organisations such as the ACLU (American Civil Liberties Union) and the EFF (Electronic Frontier Foundation) have been especially strong advocates of the widespread use of encryption since the Snowden revelations. The direct impact of these developments can be observed in the fact that many social media companies implemented stronger encryption, such as WhatsApp's end-to-end encryption. Privacy advocates try to foster the use of encryption, whereas state agencies present it as a negative tool that will prevent legitimate access to the data of, for example, criminal suspects. And while telecommunications companies often openly claim to be politically neutral, their capabilities for surveillance and data retention shape, often in covert ways, access to information and the possibilities for privacy. Indeed, one can speak of a 'surveillance–industrial complex' where 'those involved in the business of selling surveillance and security are inevitably embroiled in the politics of fear' (Hayes, 2012: 172). This industry is growing rapidly, as is the market for new security and surveillance technologies. Customers are private businesses as well as government agencies that outsource security services (Hayes, 2012: 167; see also Marquis, 2003). Encryption is one crucial part of this industry as well as an important aspect of surveillance controversies, as I will examine in detail in Chapters 3 and 4.

Privacy, surveillance and diffuse security

The brief summary above shows how encryption is embedded in debates about privacy and security. Users consider digital encryption to be a potent tool for protecting their privacy against surveillance by the state and ICT companies. Traditional definitions of privacy focus on access to information, such as Alan Westin's: 'the claim of individuals, groups or institutions to determine for themselves when, how, and to what extent information about them is communicated to others' (Westin, 1967: 3). This definition, however, runs into problems when we try to analyse contemporary monitoring practices. For example, predictive data mining can uncover beliefs and attitudes about a person that might not even be known by that person (Millar, 2009). The Cambridge Analytica scandal provides another instructive example. This company used Facebook data generated through apps that offered free personality tests, and once the scandal broke many people became aware for the first time that their 'private' data had been used for purposes beyond their control. Here, the content of the data was perhaps less of an issue than the fact that it was used for targeted advertising in the context of political campaigning. Claims to privacy acquire political significance to the extent that they challenge the power of state actors as well as private enterprises. In Chapter 5, I will examine how claims to privacy unfold in the debate around encryption. I show how claims to privacy in these controversies are tightly linked to claims about security. Privacy and security are often presented as two opposing values that need to be balanced in something like a zero-sum game (Waldron, 2003). Ultimately, however, these debates tend to prioritise one over the other (Čas *et al.*, 2017: 7). This is why claims to privacy need to be understood in the broader context of debates about surveillance and security, and indeed the legitimacy of commercial and state actors and their monitoring practices.

Both commercial and state actors monitor citizens and users in order to generate data that is used – among other things – for targeted advertising and law enforcement. This theme runs through the whole encryption debate, since encryption is one tool to disrupt these activities. More importantly on a conceptual level, these surveillance practices cannot be grasped by relying on state-centred categories that focus mainly on institutional practices and regulations (Haggerty and Ericson, 2000; Huysmans, 2016). Surveillance practices are ubiquitous: digital data is traced and collected by companies to allow for targeted advertising; CCTV cameras monitor 'public' spaces; RFID chips allow people and goods to be tracked around the world (Lyon, 2003, 2014; Nissenbaum, 2010). Security and surveillance practices are enacted through these networked, decentred technologies that transcend established boundaries of public and private and scales of local and global. Privacy can then hardly only be understood as a right secured by state institutions; rather, it is a practice entangling a variety of actors and transcending the distinctions between public and private, offline and online (Bellanova, 2017; De Goede, 2014).[4]

One way to grasp surveillance and security practices of this kind is based on acknowledging how they are dispersed and decentred. That is why it makes sense to speak of a 'diffusion' of security in a context where security practices disperse multiple insecurities and threat images rather than intensifying them and thereby creating a state of emergency (Huysmans, 2014: 87–88). This theme of diffusion and decentring recurs throughout the book. Encryption as a security technology does not provoke extraordinary measures. The security practices I discuss, such as monitoring, hacking and securing networked technology, are decentred in a way that they involve multiple actors and sites. These multiple security practices do not all culminate in one centre, but security practices entangle multiple sites and actors (Huysmans, 2014: 86). Furthermore, I argue that diffuse security practices go hand in hand with diffuse socio-technological controversies. The contestation of encryption occurs in multiple sites and is carried out over political as well as technological questions. This does not mean that political institutions become irrelevant for the politics of security technology. In Chapter 6 I discuss the role of standard-setting agencies, and in Chapter 5 we will see how law enforcement agencies are core actors in encryption debates. Focusing on the diffusion of security draws attention to the way in which established categories are transcended by new practices, but also to how, for instance, notions of national security are reaffirmed in encryption controversies. In Chapter 7 I will return to this question, arguing that the dissident discourse is often still caught up in a traditional vocabulary that hampers its ability to develop political force.

The notion of diffuse security speaks to the interest in technological controversies, which are not bound to a particular institution. Indeed, as I will show throughout Chapters 4, 5 and 6, these controversies emerge both inside and outside governmental institutions. They are diffuse in the sense that they are quite issue-specific and entangle multiple actors. In order to grasp the significance of these controversies, what is needed is a 'thick' concept of politics. By this I mean an understanding of politics that does not reduce political space to parliamentary processes or existing institutions, but instead encompasses media and the role of bureaucracies, and is more sensitive to the political character of practices outside state institutions.[5] In a context where many security practices occur outside societal institutions 'the question of democratic politics cannot by default or uncritically fall back on reiterating familiar institutional repertoires of democratic action' (Huysmans, 2016: 82). This is why my empirical investigation looks at multiple sites where encryption became a contested issue and includes controversies around security as well as technology. Attention is paid to the way in which multiple actors contest encryption and the extent to which they challenge or reaffirm 'institutional repertoires'.

I translate these conceptual choices into the notion of 'publicness', which helps me to tease out the politics of encryption controversies. Before I introduce this theme in more detail, however, some remarks on my perspective on technology are necessary.

Politics of technology

I approach political questions about security through a focus on a specific technology. Within International Political Sociology (IPS) and Critical Security Studies (CSS), there is growing interest in the political character of technology. For this book, research concerned with the politics of security technology has been especially constructive. Drones, Big Data and biometrics are just a few examples of technologies that are imperative for current security practices (Flyverbom *et al.*, 2017; Holmqvist, 2013; Jacobsen, 2015; Lyon, 2014; Sauer and Schörnig, 2012). These technologies make mass-surveillance possible or change the conduct of warfare (Cudworth and Hobden, 2015; Kroener and Neyland, 2012). A core insight here is that security practices are a result of a complex assemblage of humans, objects and actions (Jeandesboz, 2016; Leander, 2013; Salter, 2015). For instance, security practices at the European border are increasingly automated with the result that security is enacted at the border through assembling actors, technologies and devices. In this process myriad actors and technologies are brought together 'into a coherent enactment of a [future] border' (Bourne *et al.*, 2015: 308). The border and its related idea of security thus rely on the stabilisation of a complex assemblage. The political significance of security technology may thus lie in the way it enacts borders or, as I will discuss later, a certain understanding of security. For example, airport security devices enact a certain idea of security by identifying certain objects or substances, such as liquids, as per se insecure (Schouten, 2014). In this regard, I will show how uncertainty about future technological development and the validity of expert knowledge shapes ideas of security in the current encryption debates. An investigation into the politics of security technology then looks at how relations between technologies, actors and security are made and how they are stabilised. Such an 'analytics of devices' makes 'it possible to ask whether and how they [devices] make a difference in the enactment of security' (Amicelle *et al.*, 2015: 297).

In the context of security technology, such a perspective can translate into two sets of research objectives by asking: a) how the materiality of security devices creates certain kinds of security; and b) how certain technologies and their embeddedness in societal struggles make possible specific security practices and – most importantly – what this implies for politics. In this book the emphasis will be on the latter questions.[6] As I will elaborate in Chapter 2, the focus is less on the materiality of encryption (or the lack thereof) than on the politics of technological controversies.

Studying the politics of encryption works through a sensibility for the contentious nature of a technology that is deeply embedded in debates about security, surveillance and privacy. Conceptually, I develop this sensitivity into a methodology in Chapter 2. For this purpose, I will draw on Andrew Barry's work on 'material politics' (Barry, 2013). The two concepts of 'political situation' and 'technological society' are important here (Barry, 2001: ch. 1; 2012). Barry does not define a technological society by the appearance of specific technologies. The crucial change he observes takes place when societies become 'occupied' by technologies (Barry 2001: 1–10). According to Barry, we live in a technological society to 'the extent that specific technologies dominate our sense of the kinds of problems that government must address, and the solution that we must adopt' (Barry, 2001: 2). Concepts such as the network are not only used to describe technologies but also to make sense of societal phenomena such as terrorism (Friedrich, 2016). Solutions to these problems are then also formulated within this language game (Arquilla and Ronfeldt, 2001). The concept of a 'technological society' allows us to work on the distinct aspects of networked technology without losing sight of the continuities that might exist (Bonelli and Ragazzi, 2014) and the societal aspects that are equally important for a shift in security practices. In his research on the changing organisation of the US military in the context of the wars in Afghanistan and Iraq, Steve Niva goes so far as to argue 'that the central transformation enabling this shadow warfare has less to do with new technologies and more to do with new forms of social organization – namely, the increasing emergence of network forms of organization within and across the US military' (Niva, 2013: 187). This means that change does not only occur because of technological innovations; it is always embedded in societal structures.

The second key concept is that of a 'political situation' (Barry, 2012). Barry develops this concept as an extension of previous research into knowledge controversies. He emphasises that controversies cannot be properly understood in their political significance without acknowledging how a particular controversy links to other controversies and is embedded in wider struggles. This is what he calls a 'political situation', one that

> [captures] how the significance of a controversy is not so much determined by its specific focus, then, but needs to be conceived in terms of its relations to a moving field of other controversies, conflicts and events, including those that have occurred in the past and that might occur in the future.
>
> (Barry, 2012: 330)

The idea here is not to study controversies as singular events, but to understand their political significance in a larger societal and historical context. To this end, I trace the encryption controversies from the late 1980s to today (Chapter 3) and examine the various instances in which encryption is a relevant technology today (Chapter 4). Furthermore, I locate the encryption debate in wider struggles over surveillance practices and the role of the state (Chapter 5).

Publics and publicness

The concept of the public is central to most liberal and deliberative theories of democracy. Here, the presence of a public is central for making claims about the legitimacy of policy decisions. However, empirically, an ideal public sphere as depicted in these theories has never really existed (Fraser, 1990). What is more, such a concept of the public is a result of social process and has to be understood in its historicity (Geuss, 2013). Following these insights,

IR scholars have investigated how different forms of publics emerge and transgress national borders (Brem-Wilson, 2017; Frega, 2017; Porter, 2014). Rather than taking 'the public' as a given object, the analysis shifts to questions about the emergence of publics and the related changing notions of the private – and what this implies for our understanding of world politics (Best and Gheciu, 2014). Best and Gheciu put this in a nutshell:

> [W]hat counts as public gets contested and renegotiated through transformational practices [...] [W]e suggest that distinctions between public and private – just like distinctions between the national and the global, economics and security – are themselves forms of power-laden practice, which reflect and in turn shape the characteristics of the broader social context in which they are enacted.
>
> (Best and Gheciu, 2014: 15)

Notions of public and private result from social processes that always enact claims about legitimacy and authority (Herborth and Kessler, 2010). References to the existence of a public make implicit or explicit claims about the legitimacy of a political action. In the context of this book, I use the focus on publicness as a way of teasing out the political significance of technological controversies. In drawing on John Dewey's concept of the public (Dewey, 1999), I read him as an author who is critical of the reification of political institutions (Dewey, 2008). According to Dewey, the task of a public is to challenge established routines and institutions that shape democratic practices. Nancy Fraser, in this context, speaks of 'counter-publics' who contest the state and the majority public (Fraser, 1990: 116, 134; see also Dahlberg, 2007). From this perspective, not one but multiple publics may exist at the same time. These multiple publics are heterogeneous, issue-specific and have the potential to challenge established institutions and power structures.[7]

I use the notion of 'publicness' to locate political controversies outside established political institutions. As described above, this focus is necessary since the security practices that are debated in the context of encryption are often diffuse. The notion of publicness can provide a focus to the analysis of diffuse security. Focusing on the controversies about encryption not only serves as a methodological entry point, but also allows us to tease out what kind of politics is enacted. As described above, controversies and forms of publicness are always understood as parts of a wider 'political situation'. I understand a particular controversy in its context; contesting a particular technology plays out in broader societal debates on the value of security. Sensitivity to emerging forms of publicness is essential to seeing how encryption controversies and the resulting form of publicness can be 'mobilised against' the state or how it can '[serve] to constitute rather than oppose the security state' (Walters and D'Aoust, 2015: 59). To be clear, the aim is not to assess the quality of publics emerging in the context of encryption in the sense of measuring them against some standard of a 'true' public in a 'functioning' democracy. Instead, I use publicness as a concept to gain analytical leverage, with a view to examining how encryption controversies speak to debates about surveillance and privacy. Through the notion of publicness, we gain an alternative entry point into the question of what we consider to be political (Åhäll, 2016).

The idea of publicness, rather than notions such as the public sphere or public opinion, serves to highlight the multiple and heterogeneous practices in which multiple publics emerge and become politically significant. I locate political contestation in spaces where highly technological debates occur. This also means that, for me, publicness is the outcome of practices and not a pre-constituted realm. The attentiveness to publicness aids engagememt with security politics as something that is not a category that is limited to the realm of the state. In Chapter 7 I build on the findings of Chapters 4, 5 and 6 in order to engage more explicitly with the question of politics. This is achieved by investigating how encryption controversies entail specific ideas relating not only to what 'security' means but also how these conceptions rely on specific ideas about citizenship, statehood and privacy. In this way, *Crypto-Politics* works outside as well as inside the well-defined imaginary of the political.

Overview of the book

Chapters 2 and 3 engage with the conceptual and methodological foundations of the book. Chapters 4, 5 and 6 then provide empirical analyses of encryption controversies before Chapter 7 summarises the results and highlights the study's implications for our understanding of politics.

Chapter 2 situates the study in IR's engagement with the social study of science and technology. I introduce the work of Andrew Barry, especially his research into material politics, in order to highlight how controversies emerge in specific political situations. These political situations are characterised by the way in which technology can become an object of contention that brings together different sites and actors. Barry's work is especially instructive since it sensitises us to the way in which controversies combine questions about science and technology, but also politics. This does not mean that I make strong claims about the novelty of digital technology or how it shapes our current society. With the concept of technological society, the attention shifts to how societies use images stemming from science and technology to discuss political issues. Translating this conceptual engagement into a methodology, I argue for a discursive approach to the study of material politics. The interest in techno-logical controversies requires attention to the way in which particular questions about encryption become contested. This is why I analysed not only mass media depiction of encryption and security but also how experts, activists and state agents involve themselves into these controversies. While the United States has a longer history of political battles about encryption, I expand the analysis to Germany to give a more detailed picture of encryption discourses.

The chapter also presents the method of analysis developed and applied here. Drawing on insights from Grounded Theory, I used a combination of open coding and interpretation that was guided by ancillary questions. This presentation of the selection of the material and method of analysis provides an insight into how the project was conducted. Thus, the aim is not to provide a simple recipe for how to conduct research, but rather to provoke a conversation about the conduct of enquiry.

With this conceptual and methodological vocabulary in place, in Chapter 3 I proceed to a more in-depth discussion of the public as it has been discussed in Science and Technology Studies (STS). This discussion of various concepts of the public develops a more thorough link between material politics, the importance of controversies and the question of what studying the contestation of encryption can tell us about politics. I focus on the work of Michel Callon and colleagues and Noortje Marres, who, in the tradition of Actor-Network Theory (ANT), discuss how new forms of publics can emerge in relation to technological controversies (Callon and colleagues) or material objects (Marres). Their work is instructive for opening up space to engage with alternative notions about how to understand the public in the technological society. Ultimately, however, I argue that other resources are needed to engage with what I call a thicker concept of politics. This is why I turn to John Dewey and his theory of democracy. His ideas have been instructive for STS, but I show that they can be read in a way that highlights how publics always challenge established concepts of political institutions. Building on this discussion, I develop the notion of publicness as a way to highlight the multiple and heterogeneous characters of publics that are emerging.

Chapters 4, 5 and 6 engage empirically with the phenomenon of encryption, each highlighting different aspects. Chapter 4 will be especially instructive for readers without much prior knowledge of encryption. This chapter serves two purposes: first, it explains some basic technological principles to allow for a better understanding of the technology at hand; and, second, it gives an overview of the main issues to consider when discussing the politics of encryption. Hackers, civil rights organisations and security agencies have been involved in struggles over encryption in the past decades. While the empirical analysis in Chapters 5 and 6 is firmly focused on the 2010s, in Chapter 4 I describe how certain conflict lines can be traced back to the late 1980s. Some important shifts can also be identified: today, no one denies the usefulness of encryption, so the debate is not about 'pros and cons' but rather about the degree of state regulation. Indeed, the positions of actors are often ambiguous. For example, one can hardly speak of 'the state' since state agencies such as the NSA, the FBI and the government can at times be found to assess encryption in radically different ways. By introducing some background information on encryption, I show how 'security' in the context of encryption is always perceived as provisional and that experts understand encryption systems as secure only to the extent that the system has not been proved to be insecure. I return to these themes in subsequent chapters. Chapter 4 also prepares the ground for an engagement with the politics of security by emphasising the different sites in which controversies occur. We can see the 'traditional' politics when we look at how governments tried to implement certain policies. But the framework of national politics is also challenged when, for instance, encryption is 'exported' by simply uploading it and making it accessible throughout the worldwide web. Thus, Chapter 4 highlights the fact that technological and societal controversies are tightly interlinked.

Chapter 5 focuses on how encryption is debated as a security technology in order to provide greater empirical insight into encryption as a security issue. The chapter presents the encryption–security controversy in the form of two narratives. In the first narrative, encryption hampers security: the state is paranoid and threatened by its own citizens' use of strong encryption. The citizens – or potential criminals – are empowered by companies to have sole access to their own data. In the second narrative, the state is the threat as it does not provide but rather endangers security for its own citizens. Encryption figures as the best tool for improving security and preventing privacy intrusions. In this scenario, the citizen needs to be activated, since only an empowered user can take care of his or her security. As a result, an individualisation of security can be observed. Conceptually, I show how the binary of public and private does not structure these security controversies and thus does not provide much analytical leverage. Especially the role of economic actors is quite ambiguous and cannot be neatly located within the two main security narratives. Thus, trying to describe the role of companies with binaries of state/non-state or security/privacy does not really work. Instead, publicness emerges around the question of what security means, in contestations over the role of the state vis-à-vis companies and around activists trying to engage citizens in the issue of encryption.

Chapter 6 presents several instances in which encryption technology becomes contentious. Two aspects are analysed in detail. The first of these is the theory of technology that becomes visible on the level of practice. Complexity, ambiguity and the need for human input are characteristics of the theory of technology in the context of encryption. Second, debates about breaking encryption are scrutinised. I analyse not only the public controversy around Apple, the FBI and the decrypting of an iPhone, but also more technical recommendations about how to secure data against future intrusions. As a result, I show that Apple challenged the role of the state as the prime provider of security. The second part of the chapter focuses on how encryption technology is discussed in bureaucratic guidelines. Recommendations about key-length of encryption appear to be much more ambiguous and do not follow an overarching technocratic logic. Building on the insight of the previous chapter, where I argue that locating contestation within the space of 'the public' is not particularly useful or analytically valuable, Chapter 6 draws attention to new spaces of contestation.

Chapter 7 ties together the previous chapters by engaging with the question of how the notion of publicness helps us to understand the politics of security. I show the ways in which the encryption discourse transcends ideas about sovereignty and national security, but at the same time cannot quite escape this language game. While the technological features of networked technology break through national frameworks, even dissidents who oppose state surveillance reaffirm distinctions such as the one between citizens and non-citizens. In sum, the encryption discourse reaffirms the existing political vocabulary while simultaneously operating outside established political categories.

Notes

1 I use the qualification of 'radical' as defined by John Dewey, whom I discuss in more length in Chapter 3. For Dewey, 'radical' means forms of contestation that are directed against reified political institutions and envision new forms of democratic practices (Dewey 2008).
2 I use the terms 'cybersecurity', 'internet security' and 'networked security' interchangeably.
3 Indeed, cybersecurity has entailed an economic dimension since its early days (Dunn Cavelty, 2007: 47, 61). As I will detail in Chapter 5, the concept of cybersecurity always entails the economy as a referent object.
4 This observation is especially important when thinking about alternative modes of resistance. Forms of 'below-the-radar means of everyday resistance' (Amicelle *et al.*, 2015: 301) can be best analysed when looking at how these forms of resistance work inside but also outside political institutions (Gilliom, 2001). The theme of activists resisting surveillance practices by promoting the use of encryption is a theme discussed in Chapters 4 and 5.
5 The concept of 'thick' politics is very loosely inspired by Huysmans (2014: 20).
6 Hence, less attention is paid to ontological questions about the 'technologiness' of encryption or its materiality proper. In this sense I differ from so-called new materialist approaches (e.g. Barad 2003; du Plessis 2017). Bonnie Honig describes such an approach; 'This way of thinking about public things is different from (though indebted to) the varieties of vitalism and thing theory that attribute agency to things and decentre the human. Here the human remains the focus, but things have agency enough to thwart or support human plans or ambitions, and we do well to acknowledge their power and, when appropriate, to allow that power to work on us or work to lessen or augment it' (Honig, 2017: 28).
7 In Chapter 3 I suggest that Dewey sets high normative standards for such a public to exist. Since I am not interested in asking about the presence or absence of a public but more in the politics that come to the fore, I introduce the notion of publicness, which is not burdened by Dewey's criteria for a 'true' public.

Bibliography

Åhäll, L. (2016). The dance of militarisation: A feminist security studies take on 'the political'. *Critical Studies on Security* 4(2): 154–168.

Amicelle, A., Aradau, C. and Jeandesboz, J. (2015). Questioning security devices: Performativity, resistance, politics. *Security Dialogue* 46(4): 293–306.

Arquilla, J. and Ronfeldt, D. (2001). *Networks and Netwars: The Future of Terror, Crime, and Militancy*. Santa Monica, CA: Rand Corporation.

Barad, K. (2003). Posthumanist Pperformativity: Toward an understanding of how matter comes to matter. *Signs* 28(3): 801–831.

Barry, A. (2001). *Political Machines: Governing a Technological Society*. London: Athlone Press.

Barry, A. (2012). Political situations: Knowledge controversies in transnational governance. *Critical Policy Studies* 6(3): 324–336.

Barry, A. (2013). *Material Politics: Disputes along the Pipeline*. RGS-IBG Book Series. Chichester: Wiley-Blackwell.

Bellanova, R. (2017). Digital, politics, and algorithms: Governing digital data through the lens of data protection. *European Journal of Social Theory* 20(3): 329–347.

Best, J. and Gheciu, A. (2014). Theorizing the public as practices: Transformations of the public in historical context. In: Best, J. and Gheciu, A. (eds) *The Return of the Public in Global Governance*. Cambridge: Cambridge University Press, pp. 15–44.

Bonelli, L. and Ragazzi, F. (2014). Low-tech security: Files, notes, and memos as technologies of anticipation. *Security Dialogue* 45(5): 476–493.

Bourne, M., Johnson, H. and Lisle, D. (2015). Laboratizing the border: The production, translation and anticipation of security technologies. *Security Dialogue* 46(4): 307–325.

Brem-Wilson, J. (2017). La Vía Campesina and the UN Committee on World Food Security: Affected publics and institutional dynamics in the nascent transnational public sphere. *Review of International Studies* 43(2): 302–329.

Callon, M., Lascoumes, P. and Barthe, Y. (2009). *Acting in an Uncertain World: An Essay on Technical Democracy*. Cambridge, MA: MIT Press.

Čas, J.*et al.* (2017). Introduction: Surveillance, privacy and security. In: Friedewald, M.*et al.* (eds) *Surveillance, Privacy and Security Citizens' Perspectives*. Abingdon and New York: Routledge, pp. 1–12.

Cudworth, E. and Hobden, S. (2015). The posthuman way of war. *Security Dialogue* 46(6): 513–529.

Dahlberg, L. (2007). The internet, deliberative democracy, and power: Radicalizing the public sphere. *International Journal of Media and Cultural Politics* 3(1): 47–64.

De Goede, M. (2014). The politics of privacy in the age of preemptive security. *International Political Sociology* 8(1): 100–104.

Dewey, J. (1999). *The Public and Its Problems*. Athens, OH: Swallow Press.

Dewey, J. (2008) Democracy is radical. In: Boydston, J.A. (ed.) *The Collected Works of John Dewey*, Vol. 11: *The Later Works, 1925–1953*. Carbondale: Southern Illinois University Press, pp. 296–299.

Diffie, W. and Landau, S.E. (1998). *Privacy on the Line the Politics of Wiretapping and Encryption*. Cambridge, MA: MIT Press. Available at: http://search.ebscohost.com/login.aspx?direct=true&scope=site&db=nlebk&db=nlabk&AN=642 (accessed 13 March 2015).

du Plessis, G. (2017). When pathogens determine the territory: Toward a concept of non-human borders. *European Journal of International Relations* 24(2). Available at: https://journals.sagepub.com/doi/abs/10.1177/1354066117710998 (accessed 30 March 2019).

Dunn Cavelty, M. (2007). *Cyber-security and Threat Politics: US Efforts to Secure the Information Age*. CSS Studies in Security and International Relations. Abingdon and New York: Routledge.

DuPont, Q. (2016). Opinion: Why Apple isn't acting in the public's interest. *Christian Science Monitor*, 22 February. Available at: www.csmonitor.com/World/Passcode/Passcode-Voices/2016/0222/Opinion-Why-Apple-isn-t-acting-in-the-public-s-interest (accessed 3 April 2017).

Flyverbom, M., Deibert, R. and Matten, D. (2017). The governance of digital technology, Big Data, and the internet: New roles and responsibilities for business. *Business and Society* 58(1): 3–19.

Fraser, N. (1990). Rethinking the public sphere: A contribution to the critique of actually existing democracy. In: Calhoun, C. (ed.) *Habermas and the Public Sphere.* Cambridge, MA and London: MIT Press, pp. 109–142.

Frega, R. (2017). Pragmatism and democracy in a global world. *Review of International Studies* 43(4): 720–741.

Friedrich, A. (2016). Vernetzung als Modell gessellschaftlichen Wandels: Zur Begriffsgeschichte einer historischen Problemkonstellation. In: Leendertz, A. and Meteling, W. (eds) *Die neue Wirklichkeit: semantische Neuvermessungen und Politik seit den 1970er-Jahren.* Schriften des Max-Planck-Instituts für Gesellschaftsforschung Köln Band 86. Frankfurt and New York: Campus Verlag, pp. 35–62.

Geuss, R. (2013). *Privatheit: eine Genealogie.* Berlin: Suhrkamp.

Gilliom, J. (2001). *Overseers of the Poor: Surveillance, Resistance, and the Limits of Privacy.* Chicago Series in Law and Society. Chicago, IL: University of Chicago Press.

Gürses, S., Kundnani, A. and Van Hoboken, J. (2016). Crypto and empire: The contradictions of counter-surveillance advocacy. *Media, Culture and Society* 38(4): 576–590.

Haggerty, K.D. and Ericson, R.V. (2000). The surveillant assemblage. *British Journal of Sociology* 51(4): 605–622.

Hayes, B. (2012). The surveillance–industrial complex. In: Ball, K., Haggerty, K.D. and Lyon, D. (eds) *Routledge Handbook of Surveillance Studies.* New York: Routledge, pp. 164–175.

Hellegren, Z.I. (2017). A history of crypto-discourse: Encryption as a site of struggles to define internet freedom. *Internet Histories* 1(4): 285–311.

Herborth, B. and Kessler, O. (2010). The public sphere. In: *The International Studies Encyclopedia.* Hoboken, NJ: Wiley, n.p.

Herrera, G.L. (2002). The politics of bandwidth: International political implications of a global digital information network. *Review of International Studies* 28(1): 93–122.

Hintz, A. and Dencik, L. (2016). The politics of surveillance policy: UK regulatory dynamics after Snowden. *Internet Policy Review* 5(3). Available at: https://policyreview.info/articles/analysis/politics-surveillance-policy-uk-regulatory-dynamics-after-snowden (accessed 23 April 2019).

Holmqvist, C. (2013). Undoing war: War ontologies and the materiality of drone warfare. *Millennium* 41(3): 535–552.

Honig, B. (2017). *Public Things: Democracy in Disrepair.* New York: Fordham University Press.

Huysmans, J. (2011). What's in an act? On security speech acts and little security nothings. *Security Dialogue* 42(4–5): 371–383.

Huysmans, J. (2014). *Security Unbound: Enacting Democratic Limits.* Critical Issues in Global Politics. London and New York: Routledge.

Huysmans, J. (2016). Democratic curiosity in times of surveillance. *European Journal of International Security* 1(1): 73–93.

Jacobsen, K.L. (2015). Experimentation in humanitarian locations: UNHCR and biometric registration of Afghan refugees. *Security Dialogue* 46(2): 144–164.

Jeandesboz, J. (2016). Smartening border security in the European Union: An associational inquiry. *Security Dialogue* 47(4): 292–309.

Kahn, D. (1997). *The Codebreakers: The Comprehensive History of Secret Communication from Ancient Times to the Internet.* New York: Scribner's and Sons.

Kroener, I. and Neyland, D. (2012). New technologies, security and surveillance. Ball, K., Haggerty, K.D. and Lyon, D. (eds) *Routledge Handbook of Surveillance Studies.* New York: Routledge, pp. 141–148.

Kuehn, K.M. (2018). Framing mass surveillance: Analyzing New Zealand's media coverage of the early Snowden files. *Journalism* 19(3): 402–419.

Leander, A. (2013). Technological agency in the co-constitution of legal expertise and the US drone program. *Leiden Journal of International Law* 26(4): 811–831.

Levy, S. (1994). Cypher wars: Pretty good privacy gets pretty legal. Available at: http://encryption_policies.tripod.com/industry/levy_021194_pgp.htm (accessed 9 February 2015).

Lyon, D. (2003). Surveillance as social sorting: Computer codes and mobile bodies. In: Lyon, D. (ed.) *Surveillance as Social Sorting.* London: Routledge, pp. 13–30.

Lyon, D. (2014). Surveillance, Snowden, and Big Data: Capacities, consequences, critique. *Big Data and Society* 1(2). Available at: https://journals.sagepub.com/doi/full/10.1177/2053951714541861 (accessed 30 March 2019).

Marquis, G. (2003). Private security and surveillance: From the 'dossier society' to database networks. In: Lyon, D. (ed.) *Surveillance as Social Sorting: Privacy, Risk, and Digital Discrimination.* London and New York: Routledge, pp. 226–248.

May, T. (1992). *The Crypto Anarchist Manifesto.* Available at: www.activism.net/cypherpunk/crypto-anarchy.html (accessed 13 March 2015).

Millar, J. (2009). Core privacy: A problem for predictive data mining. In: Kerr, I., Steeves, V.M. and Lucock, C. (eds) *Lessons from the Identity Trail: Anonymity, Privacy and Identity in a Networked Society.* New York and Toronto: Oxford University Press, pp. 103–119.

Nissenbaum, H.F. (2010). *Privacy in Context: Technology, Policy, and the Integrity of Social Life.* Stanford, CA: Stanford Law Books. Available at: http://public.eblib.com/choice/publicfullrecord.aspx?p=483442 (accessed 9 March 2015).

Niva, S. (2013). Disappearing violence: JSOC and the Pentagon's new cartography of networked warfare. *Security Dialogue* 44(3): 185–202.

Porter, T. (2014). Constitutive public practices in a world of changing boundaries. In: Best, J. and Gheciu, A. (eds) *The Return of the Public in Global Governance.* Cambridge: Cambridge University Press, pp. 223–242.

Saco, D. (2002). *Cybering Democracy: Public Space and the Internet.* Electronic Mediations. Minneapolis: University of Minnesota Press.

Salter, M.B. (ed.) (2015). *Making Things International,* Vol. 1: *Circuits and Motion.* Minneapolis: University of Minnesota Press.

Sauer, F. and Schörnig, N. (2012). Killer drones: The 'silver bullet' of democratic warfare? *Security Dialogue* 43(4): 363–380.

Schouten, P. (2014). Security as controversy: Reassembling security at Amsterdam Airport. *Security Dialogue* 45(1): 23–42.

Stern, M. (2005). *Naming Security – Constructing Identity: 'Mayan Women' in Guatemala on the Eve of 'Peace'.* New Approaches to Conflict Analysis. Manchester and New York: Manchester University Press.

Stevens, D., Daniel, P. and Vaughan-Williams, N. (2017). *Everyday Security Threats: Perceptions, Experiences, and Consequences.* Manchester: Manchester University Press.

Waldron, J. (2003). Security and liberty: The image of balance. *Journal of Political Philosophy* 11(2): 191–210.

Walker, R.B.J. (1997). The subject of security. In Williams, M.C. and Krause, K. (eds) *Critical Security Studies: Concepts and Cases.* London: Routledge, pp. 61–82.

Walters, W. and D'Aoust, A.-M. (2015). Bringing publics into Critical Security Studies: Notes for a research strategy. *Millennium: Journal of International Studies* 44(1): 45–68.

Westin, A.F. (1967). *Privacy and Freedom*. London andNew York: The Bodley Head.

2   Researching the politics of technology

This book investigates controversies in which the contested nature of encryption comes to the fore. In this chapter I will introduce some sensitising concepts and the methodological tenets underlying the study. The focus is on researching controversies. These controversies entail two dimensions. One is technological: which technological features are necessary for security? The other is societal and concerns questions of value: what do we mean by security and who should provide it? How does it relate to other values, such as privacy and freedom? Both dimensions are present in encryption controversies and both are politically significant, as subsequent chapters will show. But how does one study the way in which encryption *technology* becomes the object of societal struggles? In this chapter I introduce concepts from the social study of science and technology that I consider to be useful. This conceptual vocabulary can situate the contestation of technological features in a wider context as well as address the question of how technological contestations are linked to debates about issues such as democracy and privacy.

Two themes will be discussed in this chapter. Conceptually, I introduce Andrew Barry's work on 'material politics', which is valuable in teasing out the way in which technology features in security discourses. Translating this into a methodology, I argue that speech is indeed still important for understanding the politics of technology. Rather than examining the material underpinnings of encryption in depth, I will focus on the broader setting in which encryption controversies are embedded within a larger security debate. This was why I turned to pragmatism and especially Grounded Theory when conducting my research. The final section describes this process in more detail and provides an insight into the process of interpretation. The next section starts by locating my research within the literature on security technology in CSS. I then discuss the notion of material politics and the concepts of 'technological society' and 'political situation'. Finally, these conceptual considerations will be translated into the actual conduct of enquiry by laying out my methodology and methods. This insight into the underlying methodology and process of research enhances discussion of the empirical results presented in the subsequent chapters.

The social study of technology

Technology and world politics

An interest in the role of technology within Security Studies can be traced back to debates about the technological imperative within Deterrence Theory (Reppy, 1990; Buzan and Hansen, 2009: 170–172). More recently, the role of biological weapons or drones, for example, has been widely discussed from various perspectives within Security Studies (Boyle, 2013; Lambeth, 1997; Rae, 2014). Traditional Security Studies discusses technology with a focus on how new technologies can be instruments for increasing security by, for instance, altering the conduct of warfare or changing power relations among states. A more explicit interest in the role of technology for security *practices* can be found in the contributions by Didier Bigo and colleagues that scrutinise the role of technology more openly (Bigo *et al.*, 2007, 2010). Their research, through a Bourdieusian framework, has shown how the field of European security professionals is organised, and they offer a sociological understanding of security. They embed their analysis of technology in a broader societal context and demonstrate, for instance, the increasing importance of databases in the daily routines of security professionals. As a result, space is opened up to study the influence of processes of automation on security practice.

Starting with a more explicit interest in *technology*, IR scholars have used insights from 'new materialist' philosophy (Austin, 2017; Coole and Frost, 2010; du Plessis, 2017), Actor-Network Theory (Agathangelou, 2016; Balzacq and Cavelty, 2016; Bellanova, 2017; Bueger and Bethke, 2014; Passoth and Rowland, 2010) or Science and Technology Studies more broadly (Abdelnour and Saeed, 2014; Barry and Walters, 2003; Jacobsen, 2015; Rothe, 2015) in order to understand the role of technology, objects and devices in world politics. These approaches vary in their epistemological and ontological assumptions, but in the context of this project two insights are imperative: the idea of scientific progress as a teleological process is challenged; and, within IR, the study of technology is embedded in an attentiveness to everyday practices, locating political practices in mundane activities and objects. I will discuss these in turn.

With regard to the notion of *progress*, in its early days, the social study of science was concerned with politicising the scientific production of science (Woolgar, 1991). This was achieved through detailed empirical studies of how scientific development was (also) influenced by social factors (Latour and Woolgar, 1986). Thereby simplistic ideas about 'truth' being the arbiter for the success of a specific scientific claim were challenged (Harding, 2002; Rouse, 1996). The key assumption is that the reason why scientific ideas become stabilised as naturalised facts cannot be meaningfully explained in categories of 'truth' or 'non-truth', but only by looking at how social factors led to the widespread acceptance of a specific fact as being true. These insights were successively used for the study of technology in order to enhance our understanding of how technology develops (Pinch and Bijker, 1987). Progress in science and technology is understood as resulting from the settling of controversies. Known as 'black-boxing', this is a result of social processes. When a technology fails or an experiment is not in line with previous established facts, this 'black box' is opened and controversies emerge. Hence, when a technology 'fails', this failure cannot be assigned to nature or objective facts (Latour, 2003: 9, 11). A technology's 'success' is not just due to its natural capacities; it is also determined by social factors. With respect to this project, this means that I start from the assumption that agreement, for instance on the question of whether an encryption system should be considered secure, is a result of socio-technological controversies. Indeed, as I will show in Chapters 5 and 6, the kind of encryption that is perceived as providing security is contentious. Settling the question of what kind of encryption provides security is a process that involves many actors, and it is riven with power dynamics. The insights of the social study of science and technology are thus important for this study to understand the societal aspects of technological controversies.

With regard to the importance of *the everyday,* within IR and more precisely the broad field of International Political Sociology, the challenge lies in researching the 'intersection between global politics and mundane lives' (De Goede, 2017: 354). Technologies and specifically processes of automation not only have an impact on daily practices but also reveal the presence of world politics in mundane objects. Michele Acuto argues against a stark distinction between global politics and the everyday and favours the concept of 'assemblage' to 'advocate a rethinking of the relationship between the international contexts of IR and the commonplace realities we all partake of in our homes' (Acuto, 2014: 346). He looks at garbage and how practices of waste disposal are not only mundane activities but 'tell us how the everyday is characterized by the constant incorporation and rejection of broader governing frameworks and wider political dynamics' (Acuto, 2014: 358). Since everyday practices and devices are embedded in international relations, scrutinising these seemingly benign practices can reveal important aspects of world politics. Encryption is a technology that is relevant when looking at internet security from a wider perspective: it is relevant for users, companies and especially activists around the world. Studying this technology thus requires looking beyond international organisations and politics between nation-states. The controversies that unfold within bureaucracies and among experts and activists are highly instructive for understanding encryption as a political technology that impacts on everyday practices, business activities and NGOs' actions. This perspective of objects and technologies then widens the empirical focus of what counts as a legitimate research object for International Relations.

A more fundamental conceptual reason for studying everyday objects and practices lies in the politics enacted through mundane security technologies. A conceptualisation of security and the political that focuses on exceptional threats and security practices confined to the realm of high politics devalues the everyday, its routines and dispersed security practices (Huysmans, 1998: 581). Opening up research on security practices that occur in everyday life thus also fulfils the function of locating politics on a different plane (Huysmans, 2014: ch. 2). The technologisation of security is not the only mechanism that leads to a diffusion of security practices, but it certainly fosters this development. For example, Big Data, CCTV cameras and RFID chips facilitate monitoring practices that are not restricted to one location (Bauman *et al.*, 2014; Lyon, 2002, 2003). The political character of these dispersed surveillance practices cannot be understood merely by looking at established institutional practices. Their political significance comes to the fore mostly in the way surveillance transgresses distinctions between, notably, the private and the public. This does not entail making a 'claim of a new political ontology', but my interest in material politics outside established institutions should be understood as 'a particular methodological move that aims at bringing out the political significance of scattered little insignificant practices, things, and relations' (Huysmans, 2016: 92). For example, the impact of automated border security cannot be grasped by tracing the decision-making process to one particular instance (Hall, 2017). In order to capture the political implications of these distributed forms of decision-making, which might not even appear as sovereign decisions (Amoore, 2013: 164), a different analytical vocabulary is necessary. Hence, this book focuses on developing a sensitivity to forms of contestation that appear in a context of diffuse security.

To conclude, studying technologised security practices broadens the *empirical* focus, but it also performs an important *political* function by including dispersed practices in political analysis (Wæver, 2011). In this chapter I focus more on the methodological implications, whereas Chapter 3 develops 'publicness' as a way to tease out the politics from an analysis of these dispersed security practices.

Material politics

Methodologically, researching the political character of technology requires a vocabulary that is sensitive to the way in which technology becomes part of politics. We need a vocabulary concerned with scrutinising technology from a social science perspective. Here, Andrew Barry's work is instructive (Barry, 2001, 2010, 2012, 2013b; Barry and Walters, 2003). His work concerns the way in which controversies 'centre on the nature of expertise, [the] trustworthiness and interests of persons who claim to be experts or reliable witnesses [and the] boundaries between what does and does not count as expertise' (Barry, 2013a: 9). As I will show in Chapters 5 and 6, these are the very questions that are contentious in the debates on encryption. In his work on pipelines, Barry (2013a) traces how the construction of a pipeline in Georgia became an issue for the British Parliament. He applies ethnographic methods, but also analyses the broader medial and institutional context in which these disputes emerge. As a result, he shows how questions about technological and scientific aspects became part of controversies. Partly this is due to the ambiguous character of the material itself: for example, the reaction of the metal to the environment is difficult to predict. Experts are aware of this ambiguous character. That a specific technical question becomes a matter of debate within a parliamentary setting is then due to the social context and cannot be traced back *only* to its technological features (Barry, 2013a: 139–151).

Barry's work is valuable for an analysis of world politics as he highlights how context plays a role in politicising technology. His concept of 'political situation' can be used to highlight the multiple and complex relationships between politics and technology in order to understand not only a particular controversy but how it plays out in the wider context as well as other controversies:[1] 'Participants in a knowledge controversy may implicitly or explicitly disagree about whether a particular controversy has any relation to other controversies that have happened in the past or that are still occurring elsewhere' (Barry, 2012: 330).

The debate about encryption in the 2010s cannot be understood without acknowledging how it is embedded in discussions on mass surveillance and how it reuses argumentative patterns developed in earlier struggles over the implementation of encryption. These controversies are thus part of a larger 'political situation' in which multiple controversies are entangled and in which multiple sites and actors will be entangled. The focus is not only on the contestation of the technology as such but also more fundamental categories, such as the question of what is actually contested, who is part of the contestation and what is considered to be legitimate knowledge. Indeed, distinctions such as nature/culture are the result of social processes; they are not natural in themselves. Barry does not define what counts as 'political'; rather, he merely highlights that knowledge controversies acquire political significance in the way they are embedded in larger societal struggles (Barry, 2012; 2013a: ch. 1). This perspective allows us to understand the kind of politics that is enacted through a specific technological arrangement.

The emphasis on the political significance of controversies and political situations speaks to an emerging interest in IPS in forms of contestation in a technologised environment (Gros *et al.*, 2017). In Chapter 6 I will discuss how there is a tendency to focus on the depoliticising effects of security technology, which fosters a technocratic logic.. However, as I argue throughout this book, such a technocratic logic is not always dominant. Indeed, as Amicelle and colleagues remark with regard to forms of opposition in a technologised environment:

> Public opposition to the use of security devices such as new police databases and terrorist watch lists is certainly a more visible type of contestation than informal, below-the-radar means of everyday resistance.
>
> (Amicelle *et al.*, 2015: 301).

Controversies make these everyday practices visible, as I will discuss in Chapter 3 when highlighting how the notion of publicness can help us foreground how controversies enact a fundamental contestation of political institutions.

A focus on political situations also nurtures a more differentiated idea of how digital technology impacts society and if this impact differs radically from previous technology. We need to be cautious when facing claims about the arrival of a novel digital era in which technology is more important than before. Technology has never been *un*important. While there are certainly features of new technology that facilitate new practices that alter society fundamentally, I do not believe that the changes we are currently observing are entirely due to new technological capabilities. Rather, the role of technology in society has changed. Here, the concept of a 'technological society' allows us situate the novelty at the level of the role technology occupies in today's society. Barry defines the technological society as the extent to which 'specific technologies dominate our sense of the kinds of problems that government and politics must address, and the solutions that we must adopt. A technological society is one which takes technical change to be the model for political intervention' (Barry, 2001, 2). He locates the distinct character of technology not within the 'novel' characteristics of digital technology. Technology is not only political in the sense that it furthers political aims; rather, it becomes the object of political debates. Science and technology become foils for the perception, framing and discussion of societal problems. Barry goes so far as to say that '[a]ny attempt to contest or challenge the social order may then involve – and probably will involve – an effort to contest the development and deployment of technology as well' (Barry, 2001: 9). Indeed, in a technological society, the 'technical skills, capacities and knowledge of the individual citizen' become a core concern (Barry, 2001: 3). Current societies are occupied with technology, and technology appears to be a problem for governance while also providing the resources to solve societal problems (Barry, 2001: chs. 7, 8; Barry and Walters, 2003).

The concept of 'network' helps to illustrate the notion of a technological society. In the 1970s and 1980s this concept was used primarily to describe the organisation of society, but since the turn of the century it has increasingly been used in reference to infrastructure (Friedrich, 2016; Barry, 2001: ch. 4). Here, it not only serves the purpose of describing the design of technology (as having a specific structure); in addition, the network concept is used to describe threats and counter-measures. For instance, catastrophes might occur if highly connected infrastructures break down (Friedrich, 2016: 45, 50). This connection is especially pertinent in the discussion of 'netwar'. Here, ideas relating to the networked configurations of terrorist organisations, networked technology (the internet) and counter-measures that need to take into account these decentralised structures all come together (Arquilla and Ronfeldt, 2001). 'Network' not only serves as a concept for description but mobilises ideas about the structure of society, threat landscapes and the most appropriate reactions.

'Technological society' is not defined by the presence or absence of certain technologies, but by the specific ways in which technologies shape how we think about political problems and solutions. It thus becomes possible to distinguish 'between those forms of politics and governance that are especially focused on the politics of objects and those that are less so' (Barry, 2012: 329). Using this idea of a technological society also implies that, analytically, the 'material' of material politics carries weight only in relation to how it is embedded in politics. In that sense it differs from new materialist philosophy and a more ontological orientation towards materiality. The majority of this study is devoted to reconstructing exactly how it is embedded and how technology plays out in political controversies. Thus, much of the empirical analysis is devoted to reconstructing the theory of technology present in these debates (for other studies that use this perspective, see: McCarthy, 2015; Webster, 2016), rather than investigating the material structures of technology.

In sum, this approach allows me to understand technological controversies as embedded in a wider societal context – a theme on which I elaborate in Chapter 3. Barry's work and Science and Technology Studies more generally have argued that the link between a specific technology and its social effects might not be that straightforward. Rather than equating one encryption system with a specific effect, I look at the multiple ways in which controversies around encryption emerge. In Chapter 5, for instance, I show that the link between security and encryption is made in two contrasting ways: it is seen partly as a threat to and partly as a source of higher security.

Methodology

Tenets

Researching material politics requires *social* scientific methods that are able to provide new insights into the empirical material at hand but also allow for theoretical reflection (Herborth, 2010). But the question is: how should the politics of technology be studied? What kind of sources are relevant and how should we study them? In the remainder of this chapter I describe the sources on which this study is built and describe the practice of conducting the research. An additional challenge is posed by the reflexive character of the social world. Tools we use for conducting empirical research will influence or even constitute the object under research, and this again influences our theoretical reflection.[2] We therefore need to take seriously Loïc Wacquant's comment:

> [T]he array of methods used must fit the problem at hand and must constantly be reflected upon in actu, in the very movement whereby they are deployed to resolve particular questions [...] [O]ne cannot disassociate the construction of the object from the instruments of construction of the object and their critique.
>
> (Bourdieu and Wacquant, 1992: 30)

But this problem identified by Wacquant and Bourdieu is not recognised by all social scientists. From the methodological perspective of neopositivism, the question of how theory, empirics, methods and the role of the researcher are linked is less of a problem. Neopositivists who conceptualise social sciences as able – at least in principle – to follow the natural sciences see the biggest problem as securing the scientificity of research. This school of research still serves as a benchmark for debating methodology and methods (Jackson, 2011: 11). Because of this seemingly commonsensical character of neopositivism, one way to develop alternative methodologies is oriented towards following the tenets laid down by neopositivism. The main concern was to show that these alternative methods were equally rigorous and produced valid and legitimate knowledge. There was a need to prove one's own scientificity (Jackson, 2011: 3–10; Tickner, 2005). Trying to discuss alternative methodologies in relation to neopositivist standards allowed authors of new methodologies to gain legitimacy. Furthermore, using the vocabulary of a different scholarly tradition allows for better communication (Lacatus *et al.*, 2015; Schwartz-Shea and Yanow, 2012). One of the most influential attempts in this regard was Grounded Theory, developed by Anselm Strauss and Barney Glaser (Glaser and Strauss, 1965, 1967; Strauss and Corbin, 1991; Clarke, 2005). My methodology builds on these insights, as I explain below.

Research using alternative methodologies has gained ground over the last few decades, especially in European IR (see, for example, the contributions in Aradau *et al.*, 2015). As a result, another way of developing methodologies becomes possible. Rather than discussing methodology in relation to neopositivism, it is now possible to discuss different methodological standards in their own terms. This allows us to develop new sensitivities to the way in which methods make certain aspects of the social world (in)visible (Law, 2004).[3] In this regard, the rest of this chapter briefly describes the 'mess' of research to give a better insight into the conduct of enquiry. The idea is not to fulfil some pre-existing methodological standard, but to provide a more transparent account of how knowledge was generated (Aradau and Huysmans, 2014; Lobo-Guerrero, 2012; Neal, 2012). A useful metaphor to think about methods in this way is provided by Anna Leander. For her, using methods is not about following a recipe as it might be laid out in a cookbook, but in recognising that methods evolve over the course of a research project and depend on the researcher. Especially when researching objects that 'cross the categorical boundaries that usually serve to organize knowledge', a fluid approach to method is needed (Leander, 2017: 236). Importantly, the 'encyclopaedia' is unfinished and, like an online encyclopaedia, even its structure is not fixed (Leander, 2017: 235). In this way questions about the role of methods in the process of enquiry and the way those methods make certain aspects of the world visible become central.

From this perspective, the process of presenting, discussing and evaluating methods is aimed more at evaluating strengths and weaknesses than at establishing one's scientificity. Describing one's conduct of enquiry is also imperative from a research-ethical perspective. It is necessary to report on the conduct of enquiry in such a way that the research community can evaluate it. In addition, a textual analysis, such as this one, faces the problem that the aim is to produce open research. The danger is that we, as researchers, project our own ideas and concepts onto the object. Thus, crucial methodological tools include means to safeguard against wishful thinking. The method of interpretation I present (in abbreviated form) below allowed me to guard against finding merely what I expected to find.

In addition, writing about the methods used serves important functions for teaching and learning. Learning how to interpret texts is possible only if previously conducted research projects are discussed (Mutlu, 2015). If a research project is conducted for the first time, guidelines, frameworks and rules may be helpful in order to develop an independent project. They can be seen as tools for learning how research is conducted, and they allow the questioning of methodological choices (see Bendix, 2013). The aim is thus not to present my methods in a way that a textbook on method would present them. An open method section allows the interested reader to learn from it. While specific methodological tools may serve as a good set of guidelines for novices, they should not be fetishised. A balance needs to be struck between creativity and following guidelines. In this section I show how I dealt with this balancing act.

Interpreting texts

Studies conducted on material politics often rely on the study of speech, even though this is rarely reflected upon. Studies by, for instance, Bruno Latour on the production of scientific facts rely on analysing citation patterns and, in the case of Andrew Barry's work on the politics of pipelines, both ethnography and speech in the form of official documents and media articles are relevant. Texts not only provide a tool that is often easily accessible but also present prevailing patterns of thought in a solidified way. Since my interest lies less in, say, the design of encryption systems and more in the way in which encryption is embedded in security debates, analysing speeches, articles and statements is the most appropriate approach.[4]

In line with interpretative research, I am not interested in trying to access the 'real intention' – that is, the supposed mental state of a person (Yanow and Schwartz-Shea, 2006). First, this is impossible. Moreover, the intersubjective meaning, rather than the subjective intention, is of paramount importance for the social scientist (as opposed to the psychologist) (Franke and Roos, 2013). In this sense I follow the underlying ideas of discourse analysis, which in itself is a vast field relying on different metatheoretical tenets (Foucault, 1972; Keller, 2007; Wodak and Meyer, 2009). In IR, discourse analyses are now very common (e.g. Hansen, 2013; Herschinger, 2010; Milliken, 1999). Studying political discourse is seen as one of the most valuable ways of understanding current dynamics of power and politics. Through the analysis of discourse, often with a focus on specific linguistic features such as metaphors or narrative structures, it becomes possible to understand these structures of meaning. Again, the aim is not to ascribe specific intentions or motives to the author of a text. Instead, a discourse analysis is interested in understanding the shared knowledge that is present in a specific community. Analysing this shared knowledge can, for example, elucidate why specific kinds of action may have seemed legitimate, or why other actions became less likely or completely unthinkable (Hajer, 1995). The assumption of discourse research is that the prevalence of specific patterns will make particular (speech-)acts more or less likely. This allows one to understand, for instance, which threat scenarios are feasible, which patterns of justifications are utilised, and which arguments regarding the power of technology may be uttered.

My research project applies this perspective to the material. Through a textual analysis, I was able to understand the prevailing patterns of presenting encryption and security. Ultimately, I was able to reconstruct how encryption is a contested issue, the technological features that are debated and how encryption is embedded in threat narratives. The next two subsections describes the selection of material chosen for interpretation and my method of analysis.

Selection of material

It was impractical to look at the whole discourse on encryption on a global scale. Thus, in order to achieve the most relevant results, I decided to focus on just two countries: the United States and Germany. As I will describe in more detail in Chapter 3, during the 1990s encryption was keenly debated in the United States. Because of its economic power, the United States is able to shape many of the policies revolving around encryption. Since it dictates many of the dominant policies in the realm of the internet, one may assume that patterns of argumentation prevailing in the United States will feature prominently in debates in other countries, too. Anecdotal evidence from research into discourse in the United Kingdom and an analysis of the debates in Germany seem to confirm this assumption (Schulze, 2017). The United States also proved to be an interesting case since it was the source of the Edward Snowden revelations, which rekindled debates about encryption. Finally, large US companies such as Google and Facebook are dominant actors in the debates about encryption. In sum, the United States is a focal point in the debates about encryption and therefore provides an ideal case for further scrutiny.

In addition, I analysed Germany, which is considered something of a 'safe haven' for internet activists because of laws that place a greater emphasis on privacy than those of other countries. Furthermore, the self-description of Germans is similar and emphasises the value of privacy, but also presents the country as different from either the United States or the United Kingdom by emphasising freedom over security. In addition, Germany is second only to the United States as a producer of encryption software, so it would benefit from stricter regulations in the latter country. For these reasons, it was reasonable to assume that the German discourse on encryption would differ starkly from that in the United States. Yet, both countries are democratic, so I expected to find a prevalent tension between security and democracy.

Following the insights of Grounded Theory, I started with a small set of empirical material and expanded it in several rounds of empirical research. First, I looked at experts' statements during the 'Digitale Agenda' hearing in the German parliament, which took place on 7 May 2014. The documents analysed included the testimonies of Dr Sandro Gaycken (Maths and Computer Science, Free University of Berlin), Professor Niko Härting (partner in his own law firm), Pascal Kurschildgen (IT security consultant) and Linus Neumann (representative of the Chaos Computer Club). The hearing introduced the main arguments in the German debate. Since the experts often linked technical explanations with political recommendations, I gained a sense of how experts perceived technology in general and what they made of encryption as a technology. Since their statements were quite lengthy (about ten pages each), they allowed me to follow some complex lines of argument and assessments.

Next, I analysed media outlets, which made up the majority of the analysis. This focus on the media is justified by the insight that the media discourse is of crucial importance in understanding *the* societal discourse. Although we live in a functionally differentiated society, the media discourse is crucial in that it can be said to mediate between various specialised discourses (Link, 2006; 2013: 2). Analysing mass media is one of the preferred sites for discourse analysts to identify prevailing patterns of argumentation in a society (Fairclough, 1995; Meier and Wedl, 2014). The main part of the analysis focuses on the presentation of encryption in newspapers. These are prime sites for analysis as they are able to make debates in a specialised field accessible to a broader audience (Wessler *et al.*, 2008: 26–28). In the newspapers I could detect patterns of argumentation stemming from both popular culture and expert discourse (Link, 2006). Although online sources are becoming ever more important, traditional newspapers are still a dominant tool for accessing news (Nossek *et al.*, 2015). In addition, I analysed specialist magazines that focus on IT but are still read by laypeople, such as *Wired*.

Most of the material dates from June 2012 to June 2016, with supplemental material dating from June 2016 to June 2018 added to corroborate the main findings and address some recent developments. This enabled inclusion of the important controversy between Apple and the FBI as well as the debates around ransomware. I analysed texts published in two major German newspapers – *Süddeutsche Zeitung* and *Frankfurter Allgemeine Zeitung* – which are considered the leading liberal and conservative newspapers, respectively. Similarly, I chose two major US newspapers – the *New York Times* and the *Washington Post*. The newspaper and magazine texts were selected through the databank *factiva*. First, I collected texts via a keyword search (*[Verschl?*

OR Krypt?] or *[encrypt? OR crypto?]*). Next, I manually selected all of the relevant texts, since *factiva* would come up with many articles that were irrelevant to the topic (such as movie reviews). I also excluded any articles that were shorter than one page. I then did a rough analysis of the remaining texts in order to select those I would subject to closer analysis.

In addition, I included some texts by German and US activists and state actors. Here, I chose those actors who are most relevant for the debate. In Germany, these were: the Chaos Computer Club (CCC), which is not only Europe's largest hackers' organisation but also an important voice in the German debate; the activist group Netpilots (Netzpiloten); and the Crypto Group of the Society for Computer Science (Fachgruppe Krypto der Gesellschaft für Informatik). Meanwhile, German politicians were surprisingly silent on the issue. Indeed, the only detailed statement I found concerning encryption was a speech by the Secretary of the Interior at the time, Thomas de Maizière. The BKA and BND (the German Federal Police and Secret Service, respectively) provided a little material on their homepages. In the United States, by contrast, the FBI is quite vocal on the subject, so it was included in the corpus alongside a speech by President Barack Obama. Finally, I included statements by the American Civil Liberties Union (ACLU) and the Electronic Frontier Foundation (EFF) – two major representatives of civil society in the US debate.

In total, I performed fine-grained analyses of 66 primary sources (see the Appendix for full details).[5] Their length varies, but most are between four and six pages long. They were interpreted in three main rounds (March–December 2014, February–August 2016 and July–August 2018). I started the analysis of the US material after analysing most of the German material. Finally, after much of the conceptual work had evolved, I analysed the controversy around key-length (see Chapter 6).

One final point should be noted: although the material is attributed to specific actors, in practice the boundaries between actors were not well defined, and similar arguments were advanced by a variety of different actor groups. For instance, as I will explain in more detail below, arguments that were formulated by activists were also presented in newspapers and specialist magazines.

Method of analysis

As a first step, I divided all of the texts into shorter sequences. I then used 'open coding' as the first approach towards the material.[6] However, in order to zoom in on encryption and security, this open coding process was accompanied by ancillary questions that guided the analysis (Roos, 2010). These questions, which were developed before the start of the research project and adjusted after the first empirical analyses, may be divided into two categories. Those in the first category had their origins in narrative analysis (Viehöver, 2001) and help the researcher to understand and summarise the broad context of the text. Especially when analysing newspaper articles, they helped me to understand the 'plot' of the piece. They were quite general in character and allowed me to focus on the main ideas expressed in the articles as well as investigate the main actors, their characterisation and what was seen as the main problem. The questions were:

- What is considered to be the problem?
- What is the solution?
- Who are the main actors?
- How are they characterised?

The second set of questions were more concerned with particular concepts and enabled me to examine the triangle of security, politics and technology. They were also inspired by insights from narrative analysis, but were more specifically tailored to the research question. They built on prior external knowledge about the changing concept of security (ideas such as risk, precaution or widening the concept of security). I investigated how the role of the state was characterised (the focus on publicness was not part of the initial research design; it emerged only during the research process) and assessed how encryption and technology in general were presented. The questions were:

- What kind of understanding of security becomes apparent (inner/outer security, deterrence/mitigation)?
- How is the role of the state described?
- How far is the state seen as being able to provide security?
- How does the dichotomy between freedom and security play out?
- How are the state, the citizen, the economy or other actors/groups/functional systems put into this relationship?
- How is technology perceived? Positively? As a threat?
- Are the problems and solutions understood as technological or political?
- How might technologies be made safer? By whom and through what means?

The two sets of questions allowed me to understand how ideas revolving around encryption were expressed. They enabled me to focus on how encryption was characterised, if it was perceived as a problem or a solution, and how the different actors who referred to it in the texts related to the central problems. I could thereby focus on assumptions that were made about security and technology, and how they are related. Ultimately, this led to the realisation that encryption is highly contentious and consequently to the focus on publicness. Especially at the beginning of the research project, the two sets of questions were crucial in guiding my analysis and avoiding the ever-present danger of losing track of the core ideas during open coding. When I became more experienced and my ideas developed, the role of the questions decreased. For instance, once I solidified the idea that security is understood as an ever-present risk, I created the 'risk' category. When looking at subsequent sequences, I would only ask if the sequence affirmed that category or not. In this way the process of open coding becomes more structured and allows the researcher to narrow her analysis and develop more relevant categories.

The interpretation of the sequence was first written by hand; then the sequences and interpretations were transferred to a digital document. A memo was written for each relevant sequence. Writing down memos is crucial for developing ideas that will lead to an answer to the research question (Strauss and Corbin, 1991: 6; Glaser and Strauss, 1967: 101–109). These memos were rewritten, compared throughout the whole research process and never viewed as final (the constant comparison method). Their length and breadth were greater at the beginning of the research than at the end. Importantly, these notes should always contribute to the formulation of tentative arguments. They are not a mere colouring scheme; instead, they ensure that the process of writing starts during the analysis. Thereby, the development of arguments is closely tied to the material and not delegated to the very end of the research process. During the whole research process, central categories develop from the written memos. These categories refer to more than one sequence and are on a somewhat higher level of abstraction. A memo for each category was written, too. These categories were initially quite descriptive, such as 'depiction of security' or 'cyberwar', and mainly mirrored the guiding questions, but they developed throughout the project. Thus, more conceptual categories, such as 'risk' or 'the citizen as user', eventually emerged. These categories became the building blocks for the presentation of the empirical results.

In sum, I have shown the process by which my empirical analysis proceeded. The advantage was that a modest amount of material was sufficient for the research process. If each text is analysed in great detail – to a high resolution, as it were – the saturation point is reached after only a few texts. Once I felt that new texts would fail to reveal any new arguments, depictions or ideas, I ceased the analysis and started to document the results.

Conclusion: understanding encryption controversies

The politics of encryption unfolds not only within governmental institutions. In addition to state actors, mass media and activists are crucial to understanding the global politics of encryption. Hence, I introduced a conceptual and methodological vocabulary of material politics that was able to capture these multiple processes. Insights from the tradition of the social study of science and technology provide valuable resources for studying the politics of encryption. More broadly, these insights speak to CSS's and IPS's interest in mundane objects and practices (Acuto, 2014; Neumann, 2007). In this chapter I focused especially on the concept of material politics in order to highlight how controversies emerge in 'political situations'. This concept refers to a multiplicity of socio-technical controversies that gain political significance in specific societal settings. However, not all technology is understood as being political in the same way. The context and actors involved in these controversies are crucial for understanding the material politics. This was why I did not look primarily at how, for instance, encryption programs evolved, but rather at the way in which technical controversies are embedded in political controversies about security and democracy. This justifies my selection of material and the method of analysis that I articulated in this chapter. I described the underlying methodology of interpretive research and laid down my method of analysis. However, before presenting the empirical results, a more detailed engagement with 'controversies' is necessary as well as an explication of how those controversies can be theorised as aspects of politics and how a study of controversy becomes a study of *politics*. Therefore, Chapter 3 will focus on these issues through an investigation of the concept of 'publicness'.

Notes

1 This focus on context has been especially highlighted by postcolonial authors (Anderson and Adams, 2007). Excluding context and history would implicitly reinforce a specific Western narrative. Going beyond a Western-centred perspective therefore allows understanding different systems of knowledge production without again falling back on categories such as 'true' or 'false' (Verran, 2002).
2 The concept of theory and empirics should not be understood as a hard and fast distinction but more in the Rortian sense as 'flags' to mark a binary (Rorty, 1989). Throughout this study, it will become clear that the two aspects cannot be neatly separated.
3 Authors concerned with (auto-)ethnography and narrative research have been especially vocal in trying to generate different means of providing accounts of research. For a variety of standpoints, see Doty (2004), Edkins (2013), Naumes (2015) and Hamati-Ataya (2014), among others.
4 In addition, it could be argued that new materialists' emphasis on 'materiality' in particular misinterprets much of the poststructuralist and constructionist thought present in IR and STS. See, for instance: Lundborg and Vaughan-Williams (2015) and Hacking (1988).
5 Of course, this does not mean that more 'empirical' material was not used in this study. Rather, the 66 texts were analysed in this fine-grained way in order to go beyond the mere content and understand the construction of certain patterns.
6 The method of analysis relies heavily on Grounded Theory. However, I did not follow the suggested form of coding verbatim; instead, I used this theory's ideas as a guideline and relied on Roos's (2010) detailed description of his coding procedure. In addition, Ulrich Franke and Benjamin Herborth analysed material with me.

Bibliography

Abdelnour, S. and Saeed, A.M. (2014). Technologizing humanitarian space: Darfur advocacy and the rape-stove panacea. *International Political Sociology* 8(2): 145–163.

Acuto, M. (2014). Everyday International Relations: Garbage, grand designs, and mundane matters. *International Political Sociology* 8(4): 345–362.

Agathangelou, A.M. (2016). Bruno Latour and ecology politics: Poetics of failure and denial in IR. *Millennium: Journal of International Studies* 44(3): 321–347.

Amıcelle, A., Aradau, C. and Jeandesboz, J. (2015). Questioning security devices: Performativity, resistance, politics. *Security Dialogue* 46(4): 293–306.

Amoore, L. (2013). *The Politics of Possibility: Risk and Security beyond Probability.* Durham, NC, and London: Duke University Press.

Anderson, W. and Adams, V. (2007). Pramoedya's chickens: Postcolonial studies of technoscience. In Hackett, E.J.*et al.* (eds) *The Handbook of Science and Technology Studies.* London: MIT Press, pp. 181–204.

Aradau, C. and Huysmans, J. (2014). Critical methods in International Relations: The politics of techniques, devices and acts. *European Journal of International Relations* 20(3): 596–619.

Aradau, C. *et al.* (eds) (2015). *Critical Security Methods: New Frameworks for Analysis.* New International Relations Studies. London: Routledge.

Arquilla, J. and Ronfeldt, D. (2001) *Networks and Netwars: The Future of Terror, Crime, and Militancy.* Santa Monica, CA: Rand Corporation.

Austin, J.L. (2017). We have never been civilized: Torture and the materiality of world political binaries. *European Journal of International Relations* 23(1): 49–73.

Balzacq, T. and Cavelty, M.D. (2016). A theory of actor–network for cyber-security. *European Journal of International Security* 1(2): 176–198.

Barry, A. (2001). *Political Machines: Governing a Technological Society.* London: Athlone Press.

Barry, A. (2010). Materialist politics. In: Braun, B. and Whatmore, S.J. (eds) *Political Matter. Technoscience, Democracy, and Public Life.* Minneapolis: University of Minnesota Press, pp. 89–118.

Barry, A. (2012). Political situations: Knowledge controversies in transnational governance. *Critical Policy Studies* 6(3): 324–336.

Barry, A. (2013a). *Material Politics: Disputes along the Pipeline.* RGS-IBG Book Series. Chichester: Wiley-Blackwell.

Barry, A. (2013b). The translation zone: Between Actor-Network Theory and International Relations. *Millennium: Journal of International Studies* 41(3): 413–429.

Barry, A. and Walters, W. (2003). From EURATOM to 'complex systems': Technology and European government. *Alternatives: Global, Local, Political* 28(3): 305–329.

Bauman, Z.*et al.* (2014). After Snowden: Rethinking the impact of surveillance. *International Political Sociology* 8(2): 121–144.

Bellanova, R. (2017). Digital, politics, and algorithms: Governing digital data through the lens of data protection. *European Journal of Social Theory* 20(3): 329–347.

Bendix, D. (2013). Auf den Spuren kolonialer Macht: Eine genealogische Dispositivanalyse von Entwicklungspolitik. In: Franke, U. and Roos, U. (eds) *Rekonstruktive Methoden der Weltpolitikforschung.* Baden-Baden: Nomos, pp. 181–218.

Bigo, D.*et al.* (2007). Mapping the field of the EU internal security agencies. In: Bigo, D.*et al.* (eds) *The Field of the EU Internal Security Agencies.* Louvain: Centre d'études sur les conflits/l'Harmattan, pp. 5–66.

Bigo, D., Bondıtı, P. and Olsson, C. (2010). Mapping the European field of security professionals. In: Bigo, D. *et al.* (eds) *Europe's 21st Century Challenge: Delivering Liberty.* Farnham and Burlington, VT: Ashgate, pp. 49–63.

Bourdieu, P. and Wacquant, L.J.D. (1992). *An Invitation to Reflexive Sociology.* Cambridge: Polity Press.

Boyle, M.J. (2013). The costs and consequences of drone warfare. *International Affairs* 89(1): 1–29.

Bueger, C. and Bethke, F. (2014). Actor-networking the 'failed state': An enquiry into the life of concepts. *Journal of International Relations and Development* 17(1): 30–60.

Buzan, B. and Hansen, L. (2009). *The Evolution of International Security Studies.* Cambridge and New York: Cambridge University Press.

Clarke, A. (2005). *Situational Analysis: Grounded Theory after the Postmodern Turn.* Thusand Oaks, CA: Sage.

Coole, D.H. and Frost, S. (eds) (2010). *New Materialisms: Ontology, Agency, and Politics.* Durham, NC, and London: Duke University Press.

De Goede, M. (2017). Afterword: Transversal politics. In: Guillaume, X. and Bilgin, P. (eds) *Routledge Handbook of International Political Sociology.* Abingdon: Routledge, pp. 353–365.

Doty, R.L. (2004). Maladies of our souls: Identity and voice in the writing of academic international relations. *Cambridge Review of International Affairs* 17(2): 377–392.

du Plessis, G. (2018). When pathogens determine the territory: Toward a concept of non-human borders. *European Journal of International Relations* 24(2): 391–413. Available at: https://journals.sagepub.com/ (accessed 30 March 2019).

Edkins, J. (2013). Novel writing in international relations: Openings for a creative practice. *Security Dialogue* 44(4): 281–297.

Fairclough, N. (1995). *Media Discourse.* London: Arnold.

Foucault, M. (1972). *The Archaeology of Knowledge.* Translated by Sheridan Smith. New York: Pantheon Books.

Franke, U. and Roos, U. (2013). Einleitung: Zu den Begriffen 'Weltpolitik' und 'Rekonstruktion'. In: Franke, U. and Roos, U. (eds) *Rekonstruktive Methoden der Weltpolitikforschung: Anwendungsbeispiele Und Entwicklungstendenzen.* Baden-Baden: Nomos, pp. 7–29.

Friedrich, A. (2016). Vernetzung als Modell gessellschaftlichen Wandels: Zur Begriffsgeschichte einer historischen Problemkonstellation. In: Leendertz, A. and Meteling, W. (eds) *Die neue Wirklichkeit: Semantische Neuvermessungen und Politik seit den 1970er-Jahren.* Schriften des Max-Planck-Instituts für Gesellschaftsforschung Köln Band 86. Frankfurt and New York: Campus Verlag, pp. 35–62.

Glaser, B.G. and Strauss, A.L. (1965). *Awareness of Dying.* Observations. Chicago, IL: Aldine.

Glaser, B.G. and Strauss, A.L. (1967). *The Discovery of Grounded Theory: Strategies for Qualitative Research.* Chicago, IL: Aldine.

Gros, V., de Goede, M. and İşleyen, B. (2017). The Snowden Files made public: A material politics of contesting surveillance. *International Political Sociology* 11(1): 73–89.

Hacking, I. (1988). The participant irrealist at large in the laboratory. *British Journal for the Philosophy of Science* 39(3): 277–294.

Hajer, M.A. (1995). *The Politics of Environmental Discourse: Ecological Modernization and the Policy Process.* Oxford: Clarendon Press.

Hall, A. (2017). Decisions at the data border: Discretion, discernment and security. *Security Dialogue* 48(6): 488–504.

Hamati-Ataya, I. (2014). Transcending objectivism, subjectivism, and the knowledge in-between: The subject in/of 'strong reflexivity'. *Review of International Studies* 40 (1): 153–175.

Hansen, L. (2013). *Security as Practice: Discourse Analysis and the Bosnian War.* Abingdon: Routledge.

Harding, S.G. (2002). *Is Science Multicultural? Postcolonialisms, Feminisms, and Epistemologies.* Bloomington: Indiana University Press.

Herborth, B. (2010). Rekonstruktive Forschungslogik. In: Masala, C., Sauer, F. and Wilhelm, A. (eds) *Handbuch der Internationalen Politik.* Wiesbaden: VS Verlag für Sozialwissenschaften, pp. 265–284.

Herschinger, E. (2010). *Constructing Global Enemies: Hegemony and Identity in International Discourses on Terrorism and Drug Prohibition.* London and New York: Routledge.

Huysmans, J. (1998). The Question of the Limit: Desecuritisation and the aesthetics of horror in political realism. *Millennium: Journal of International Studies* 27(3): 569–589.

Huysmans, J. (2014). *Security Unbound: Enacting Democratic Limits.* Critical Issues in Global Politics. London and New York: Routledge.

Huysmans, J. (2016). Democratic curiosity in times of surveillance. *European Journal of International Security* 1(1): 73–93.

Jackson, P.T. (2011). *The Conduct of Inquiry in International Relations: Philosophy of Science and Its Implications for the Study of World Politics.* New International Relations. London: Routledge.

Jacobsen, K.L. (2015). Experimentation in humanitarian locations: UNHCR and biometric registration of Afghan refugees. *Security Dialogue* 46(2): 144–164.

Keller, R. (2007). *Diskursforschung: Eine Einführung für SozialwissenschaftlerInnen.* Wiesbaden: VS Verlag für Sozialwissenschaften.

Lacatus, C., Schade, D. and Yao, Y. (2015). Quo vadis IR: Method, methodology and innovation. *Millennium* 43(3): 767–778.

Lambeth, B.S. (1997). The technology revolution in air warfare. *Survival* 39(1): 65–83.

Latour, B, (2003). *Science in Action: How to Follow Scientists and Engineers through Society.* Cambridge, MA: Harvard University Press.

Latour, B. and Woolgar, S. (1986). *Laboratory Life: The Construction of Scientific Facts.* Princeton, NJ: Princeton University Press.

Law, J. (2004). *After Method: Mess in Social Science Research.* International Library of Sociology. London and New York: Routledge.

Leander, A. (2017). From cookbooks to encyclopaedias in the making: Methodological perspectives for research of non-state actors and processes. In: Kruck, A. and Schneiker, A. (eds) *Methodological Approaches for Studying Non-state Actors in International Security: Theory and Practice.* London: Routledge, pp. 231–240.

Link, J. (2006). Diskursanalyse unter besonderer Berücksichtigung von Interdiskurs und Kollektivsymbolik. *Handbuch Sozialwissenschaftliche Diskursanalyse* 1: 433–458.

Link, J. (2013). Diskurs, Interdiskurs, Kollektivsymbolik. *Zeitschrift für Diskursforschung* 1(1): 7–23.

Lobo-Guerrero, L. (2012). Wondering as a research attitude. In: Salter, M.B. and Mutlu, C.E. (eds) *Research Methods in Critical Security Studies: An Introduction.* New York: Routledge, pp. 25–28.

Lundborg, T. and Vaughan-Williams, N. (2015). New materialisms, discourse analysis, and International Relations: A radical intertextual approach. *Review of International Studies* 41(1): 3–25.

Lyon, D. (2002). Everyday surveillance: Personal data and social classifications. *Information, Communication and Society* 5(2): 242–257.

Lyon, D. (2003). Surveillance as social sorting: Computer codes and mobile bodies. In: Lyon, D. (ed.) *Surveillance as Social Sorting: Privacy, Risk, and Digital Discrimination*. London and New York: Routledge, pp. 13–30.

McCarthy, D.R. (2015). *Power, Information Technology, and International Relations Theory: The Power and Politics of US Foreign Policy and the Internet*. Palgrave Studies in International Relations. New York: Palgrave Macmillan.

Meier, S. and Wedl, J. (2014). Von der Medienvergessenheit der Diskursanalyse. In: Angermuller, J. *et al.* (eds) *Diskursforschung: Ein Interdisziplinäres Handbuch*. Bielefeld: Transcript, pp. 411–435.

Milliken, J. (1999). The Study of Discourse in International Relations: A critique of research and methods. *European Journal of International Relations* 5(2): 225–254.

Mutlu, C.E. (2015). How (not) to disappear completely: Pedagogical potential of research methods in International Relations. *Millennium* 43(3): 931–941.

Naumes, S. (2015). Is all 'I' IR? *Millennium* 43(3): 820–832.

Neal, A. (2012.) Empiricism without positivism: King Lear and Critical Security Studies. In: Salter, M.B. and Mutlu, C.E. (eds) *Research Methods in Critical Security Studies: An Introduction*. New York: Routledge, pp. 42–45.

Neumann, I.B. (2007). 'A speech that the entire ministry may stand for,' or: Why diplomats never produce anything new. *International Political Sociology* 1(2): 183–200.

Nossek, H., Adoni, H. and Nimrod, G. (2015). Media audiences: Is print really dying? The state of print media use in Europe. *International Journal of Communication* 9: 21.

Passoth, J.-H. and Rowland, N.J. (2010). Actor-network state: Integrating Actor-Network Theory and State Theory. *International Sociology* 25(6): 818–841.

Pinch, T. and Bijker, W.E. (1987). The social construction of facts and artifacts: Or how the sociology of science and the sociology of technology might benefit each other. In: Bijker, W.E., Hughes, T.P. and Pinch, T. (eds) *The Social Construction of Technological Systems: New Directions in the Sociology and History of Technology*. Cambridge, MA: MIT Press, pp. 11–44.

Rae, J.D. (2014). *Analyzing the Drone Debates: Targeted Killing, Remote Warfare, and Military Technology*. New York: Palgrave Macmillan.

Reppy, J. (1990). The technological imperative in strategic thought. *Journal of Peace Research* 27(1): 101–106.

Roos, U. (2010). *Deutsche Außenpolitik, eine Rekonstruktion der grundlegenden Handlungsregeln*. Wiesbaden: Springer Fachmedien.

Rorty, R. (1989). *Contingency, Irony, and Solidarity*. Cambridge and New York: Cambridge University Press.

Rothe, D. (2015). Von weitem sieht man besser. *Zeitschrift für Internationale Beziehungen* 22(2): 97–124.

Rouse, J. (1996). *Engaging Science: How to Understand Its Practices Philosophically*. Ithaca, NY: Cornell University Press.

Schulze, M. (2017). Clipper meets Apple vs. FBI: A comparison of the cryptography discourses from 1993 and 2016. *Media and Communication* 5(1): 54.

Schwartz-Shea, P. and Yanow, D. (2012). *Interpretive Research Design: Concepts and Processes*. Routledge Series on Interpretive Methods. New York: Routledge.

Strauss, A.L. and Corbin, J.M. (1991). *Basics of Qualitative Research: Grounded Theory Procedures and Techniques*. Newbury Park, CA: Sage.

Tickner, J.A. (2005). What is your research program? Some feminist answers to International Relations methodological questions. *International Studies Quarterly* 49(1): 1–22.

Verran, H. (2002). A postcolonial moment in Science Studies: Alternative firing regimes of environmental scientists and Aboriginal landowners. *Social Studies of Science* 32(5–6): 729–762.

Viehöver, W. (2001). Diskurse als Narrationen. In: Keller, R.*et al.* (eds) *Handbuch Sozialwissenschaftliche Diskursanalyse*. Wiesbaden: VS Verlag für Sozialwissenschaften, pp. 177–206.

Wæver, O. (2011). Politics, security, theory. *Security Dialogue* 42(4–5): 465–480.

Webster, M.D. (2016). Examining philosophy of technology using Grounded Theory methods. *Forum Qualitative Sozialforschung/Forum: Qualitative Social Research* 17(2). Available at: www.qualitative-research.net/index.php/fqs/article/view/2481/3948 (accessed 30 March 2019).

Wessler, H.*et al.* (2008). *Transnationalization of Public Spheres*. Transformations of the State. Basingstoke and New York: Palgrave Macmillan.

Wodak, R. and Meyer, M. (eds) (2009). *Methods of Critical Discourse Analysis*. 2nd edition. London and Thousand Oaks, CA: Sage.

Woolgar, S. (1991). The turn to technology in social studies of science. *Science, Technology, and Human Values* 16(1): 20–50.

Yanow, D. and Schwartz-Shea, P. (eds) (2006). *Interpretation and Method: Empirical Research Methods and the Interpretive Turn*. Armonk, NY: M.E. Sharpe.

3 On publics in the technological society

The previous chapter introduced the concept of material politics and offered some methodological insights into how to study the politics of encryption. The task in this chapter is to develop a more thorough link between material politics and the importance of controversies, as discussed in the previous chapter. Indeed, the question is what the study of the encryption controversies can tell us about politics. This link is made by the concept of publicness. I will show how a sensitivity to emerging publicness helps to clarify not only how controversies are crucial for democratic politics but also how these modes of publicness can be conceptualised as a fundamental challenge to existing political structures. This chapter thus also lays the groundwork for Chapter 7, in which I return to the question of how encryption controversies challenge dominant ideas about politics.

This move reflects concerns by other authors interested in the *politics* of publics (Dean, 2001; Young, 2001). Within IR, more recent contributions have engaged with the 'return of the public' (Best and Gheciu, 2014b). Acknowledging the multiple character of publics allows us to analyse how a public can emerge beyond the nation state (Brem-Wilson, 2017) and as a result of practices (Best and Gheciu, 2014a; Walters and D'Aoust, 2015). This perspective stands in contrast to concepts of the public that rely on a stark distinction between the public and the private, treating them de facto as pre-constituted realms. What counts as public and private is a product of historical formation (Fraser, 1990; Geuss, 2013). That is why it is necessary to understand any emergent forms of publicness as a result of social practices. In modern political theory, the distinction between public and private is also deeply gendered, so it not only reproduces hierarchies but also presents certain issues as private and thus non-political (Elshtain, 1981). In the context of security this gendered notion is especially powerful politically, since it carries with it a notion of which subjects and actions are relevant for security politics. Only that which is 'public' becomes visible, whereas 'private' matters are seen as irrelevant for (national) security. As I will argue throughout this chapter, modes of publicness, even though they emerge outside the established institutional landscape, still acquire political significance. The boundary between public and private is thereby challenged.

Notions of the public, of contestation and controversy also emerge in debates about the depoliticising character of security. In this context depoliticisation means a lack of public controversies. Security is considered to be depoliticising because claims of an existential threat move security measures outside democratic procedures (Buzan *et al.*, 1998; Hegemann and Kahl, 2016; Huysmans, 1998). More diffuse security practices that do not evoke the state of emergence are said to have a depoliticising effect because they either empower security professionals or invoke a technocratic logic. In both cases depoliticisation occurs through the diffusion of decision-making processes and a prevailing technocratic logic (Balzacq, 2015; Bigo *et al.*, 2010; Huysmans, 2014: 148–154). However, in this study, I am more interested in the ways in which controversies around security technology can lead to new modes of publicness. The notion of publicness that I introduce in this chapter is not meant as an assessment of the extent to which a certain 'audience' legitimises security politics (Balzacq, 2005). Instead, the aim is to find instances where not only depoliticisation but also politicisation occur. Publicness, as it is understood here, highlights the multiple, heterogeneous and possibly radical character of contestation. It is not to be understood as synonymous with an already constituted audience that legitimises or delegitimises particular policies.[1]

Throughout this chapter I will show how controversies are sometimes conceptualised as a rather tame tool, aimed merely at translating forms of contestation back into the political system. This is why I argue that we need thicker concepts of politics and the public to highlight their potential for fundamental contestation. By a thick concept of politics, I mean a concept of the political that is broad and does not constrain the political to a particular set of institutions, and which displays awareness for the broader societal context and the formation of new subject positions. It resembles the discussion of Andrew Barry's concept of political situations in the previous chapter, where Barry similarly advocates for an awareness of the context in which controversies emerge. I will draw on the work of John Dewey, which has been taken up within STS and influences much of the discussion on democracy in this field (Latour, 2005). With Dewey it becomes possible to locate the emergence of a form of publicness that is not linked back to the nation-state. In this way it becomes possible to combine STS's insights on controversies, Dewey's concept of publics and a thick concept of politics.

I will focus on two major contributions that explicitly deal with material politics, publics and democracy: the work of Michel Callon, Pierre Lascoumes and Yannick Barthe on hybrid forums, followed by Noortje Marres's engagement with Dewey's concept of the public and her work on material participation. Marres introduces American pragmatism to the debate on democracy and technology. In my reconstruction, I show that Dewey offers theoretical resources that are better suited to discussing the politics of encryption by delinking publics from national politics. The final section introduces the notion of publicness as a way to use Dewey's insights without narrowing the question to an assessment of whether we can observe the emergence of a 'real' public. In this sense, publicness acts as an analytical tool rather than a tool for normative prescription.

Hybrid forums and technological controversies

Scholars concerned with the 'public understanding of science' (PUS) have lately been concerned with the question of how to rethink democratic practices in the technological society (Stilgoe *et al.*, 2014). This strand of STS explores how citizens might be more involved in policies that revolve around complex scientific issues. PUS has been particularly interested in the democratisation of science politics and the complex relationships between state, science and citizens (Irwin, 2014; Leach *et al.*, 2007; Wynne, 2006). The PUS literature has received little attention from IR scholars, but consideration of it is necessary for my engagement with publicness in the context of security. This literature engages with ways in which 'the public' may be more involved in decision-making processes. In order to counter the power of experts and democratise decision-making, PUS research has offered valuable insights. This research has been quite influential in introducing participatory forums, and today ever more governments are concerned with involving the public (Elam and Bertilsson, 2003; Jasanoff, 2003). We can see how forms of public are issue-specific and how they bring to the fore societal problems arising because of new scientific and technological developments. Ultimately, I will show that such a conception that focuses on participation relies on a rather thin concept of politics. This literature will thus be discussed in order to shed light on the particular concept of politics on which these engagement practices rely.

I will mainly discuss Callon and colleagues' *Acting in an Uncertain World* (Callon *et al.*, 2009), which is probably the most thorough engagement with new forms of public participation and its relation to democracy. Their central contribution is the introduction of the idea of 'hybrid forums' as a way in which experts and citizens can explore complex issues together. In true STS fashion, they do not start with a theoretical engagement with contemporary theories of democracy; instead, they empirically describe various instances when technological issues became concerns for politics. Their work argues from case studies, and stories about public engagement highlight some issues that are central to the whole book. They show that complex technological issues need to be decided by and are often delegated to experts, yet citizens mistrust experts and their decisions (Callon *et al.*, 2009: ch. 1; see also Irwin, 2001; Wynne 2006).

Callon and colleagues, in line with the general outlook of social studies of knowledge, problematise the idea that academic research is the only legitimate form of knowledge production (Callon *et al.*, 2009: ch. 2). They term traditional academic knowledge production 'secluded research'. Through a brief historical sketch, they explain how academic knowledge production gained its legitimacy from the fact that it was delegated to a few experts conducting research in detached laboratories (Callon *et al.*, 2009: ch. 2). They build on research in STS and the philosophy of science on knowledge production. The result of scientific research is thus also (but not only) a social process. For Callon and colleagues, it is important to emphasise that science is often seen as a practice that is detached from the everyday. This detachment does not increase the objectivity of research but instead causes problems. Since the modern understanding of science assumes that true knowledge is gained by conducting experiments in a clean, detached environment, several kinds of 'translation' need to take place in order for this secluded research to work in the 'real world' (Callon *et al.*, 2009: 68–70). First, phenomena and problems occurring in the real world need to be translated into an experimental setting. The second kind of translation occurs within the research community when experiments are conducted, results communicated and facts black-boxed. Ultimately, these results have to be translated back in order to have an impact in the world outside the laboratory. According to Callon and colleagues, these translations may lead to problems, since the 'facts' generated in secluded research might not lead to applicable knowledge. This also implies that expert knowledge is not as apolitical or neutral as it is often presented.

Since this idea of conducting secluded research runs into problems, the authors propose a different kind of research. Although they do not dismiss the idea of secluded research altogether, they think that additional ways of knowledge production are necessary (Callon *et al.*, 2009: 83–89). They advocate for a way of research that is not detached from the world and does not delegate knowledge production to a few experts. This alternative is called 'research in the wild' (see Callon *et al.*, 2009: 74–82). While authors such as Sandra Harding and Joseph Rouse suggest that the deeply political character of science is camouflaged by the prevailing image of academia as a place for neutral knowledge production (Harding, 2002; Rouse, 1996), for Callon and colleagues secluded research is problematic for another reason. According to them, secluded research prevents the emergence of 'socio-technical controversies' (Callon *et al.*, 2009: 28). Since social and technical uncertainties exist, the emergence of 'controversies' allows exploration of these uncertainties. In controversies, future 'possible worlds' can be explored. For the authors, then, controversies are crucial for democracy since they open up space for debate and contestation about the future of a community. These controversies encompass debates over technical questions but also the formation of new communities (if, for example, a group is affected by specific technological developments). Two dimensions are explored: the possible actors who make up society (broadly understood) and the possible future states of the world (the technical dimension). What is more, in these controversies, the boundary between these two dimensions may be challenged:

> One of the central things at issue in these controversies is precisely establishing a clear and widely accepted border between what is considered to be unquestionably technical and what is recognized as unquestionably social.
>
> (Callon *et al.*, 2009: 24–25).

A public debate arises once affected people become aware of the societal dimension of a particular problem (Callon *et al.*, 2009: 25). Controversies, in contrast to secluded research, facilitate engagement with the social dimension of technical problems. If the exploration of new worlds is done in an unconstrained way, 'overflows' can emerge that 'give rise to unexpected problems by giving prominence to unforeseen effects' (Callon *et al.*, 2009: 28).

As a result, it is evident that controversies are inextricably linked with ideas about political life and the construction of policies that are able to cope with complex technological and societal questions. For Callon and colleagues, a distinct means of knowledge production is necessary to improve decision-making. This becomes clear in their discussions of instances when non-experts have become influential (Callon *et al.*, 2009: 17, 33, 76–83). Here, laypeople produced unique knowledge and influenced policy-making and scientific research. People who were affected by a common issue, such as pollution or a particular disease, came together, formed a group and tried to address the problem. In such situations new communities emerge, new identities are formed and better understanding of technological or ecological problems is achieved. Importantly, research in the wild, exploring new phenomena, not only widens the view of possible worlds in the sense of new technological developments but also allows for the emergence of new identities through the emergence of new (technical) problems and their exploration (Callon *et al.*, 2009: 146, 147).

With this groundwork in place, Callon and colleagues engage more directly with democracy. Their core concern is with how to encourage the emergence of controversies. Precisely because new states of the world, technological developments and identities can be explored within controversies, they are deemed essential for democratic politics.[2] Traditional ideas of representation must be rethought. Mirroring the problematisation of the delegation of science to experts, the idea of representation by a few professional politicians is similarly criticised. This delegation to experts prevents new identities from forming, as do traditional democratic procedures, such as elections and referendums (Callon *et al.*, 2009: 108, 115). Instead of this form of 'delegative democracy', Callon and colleagues propose a 'dialogic democracy' in which controversies will flourish. They try to open up space for the engagement of citizens and laypersons alike (Callon *et al.*, 2009: ch. 5). Considered to be crucial in this context is the concept of hybrid forums, which are, according to the authors, the best way in which to build bridges between different forms of research and enable new identities to form. These hybrid forums may take various forms, and Callon and colleagues repeatedly emphasise that their structure and procedures must remain open for negotiation, redefinition and change over time (Callon *et al.*, 2009: 118, 188).

The authors present several examples of these forums, most notably consensus conferences (Callon *et al.*, 2009: 163). These and similar forums have already been established in recent decades in several countries. Citizens come together and discuss a certain technological or scientific issue that is considered to be problematic, such as genetically modified food or climate change (Fischer, 2000; Jasanoff, 2005; Lezaun and Soneryd, 2007). During the conference, citizens are first provided with information about the issue by several experts from different areas. The aim is to present the problem in all its complexity. The citizens can then ask questions and deliberate. Finally, the results are presented at a public event, such as a press conference. Ideally, these results should influence policy-making, but this is not always the case (Callon *et al.*, 2009: 169–171). Hybrid forums are thus an expansion of democratic features that is more sensitive to controversies but still locates democratic procedure within the 'public' processes of a national society that ideally influences the behaviour of elected politicians.

The procedure described above is considered desirable by Callon and colleagues for normative reasons: they will always promote (social) learning and are democratic since they are considered by participants to be fair and to provide just results (see Callon *et al.*, 2009: ch. 7). The authors therefore assume that hybrid forums are a way to improve collective binding decisions over time. Although STS scholars traditionally treat the idea of progress (technological or scientific) in its naive form with great scepticism, in their discussion of hybrid forums Callon and colleagues are surprisingly optimistic about the possibility of collective learning that will only ever be positive (Callon *et al.*, 2009: 32, 34). This is remarkable, given that the authors start their argument with a critical account of knowledge production and show why these processes are never as neutral as one might assume. However, they do not seem to see the possibility that hybrid forums might ever lead to normative, non-desirable decisions. The problem of how to protect minorities, for example, is only briefly discussed, and the assumption seems to be that in hybrid forums the threat of the likes of sexism or xenophobia is only a side issue. Previous consensus conferences are assessed as overwhelmingly positive, with the few negative features dismissed as *procedural* deficits (e.g. a lack of binding decisions). The concept of hybrid forums, the way they are set up and the wider societal implications are not seen as possible sources of negative effects. Callon and colleagues focus on the practice and thus exclude more in-depth reflections on the particular politics enacted through hybrid forums.

Another question arises when looking more closely at the relationship between dialogic and delegative democracy. For the authors, dialogic democracy is an addition to current forms of representative democracy. Classic democratic institutions are not challenged but ought to coexist with hybrid forums (Callon *et al.*, 2009: 253). Hybrid forums, however, should increasingly become a central part of democratic deliberations. Callon and colleagues do not explicitly state in which cases representatives should decide. They insist, however, that time-sensitive issues (as security is traditionally understood) should not be excluded from hybrid forums (Callon *et al.*, 2009: 222). A more fundamental problem is their lack of recognition of issues such as the possibility of manipulation and framing, and subtle forms of the exercise of power are not discussed thoroughly. They present some caveats towards the end of their final chapter, but possible criticisms are not convincingly addressed. One such point is that procedures are not as open to dissident and marginalised voices as they might seem (Callon *et al.*, 2009: 248, 249). Callon *et al.* (2009: 249) counter this by stating that this critique underestimates the power of these 'marginalised' actors:

> To play down the influence of the actors' cynicism and tactical skill is not to sin by excessively naive optimism. Yes, the actors are calculators, cynics, and Machiavellian. But tactical skill is not the monopoly of any one group of actors.

Thus, all of the actors enter the debate with the same resources (tactical skills) and all are able to manipulate the debate. Since the procedures will continuously reform, they will become more dialogic over time (Callon *et al.*, 2009: 249). This assessment does in fact strike me as rather naive. Once a hybrid forum is set up, with free discussion taking place, STS's insights concerning the linkage between power and knowledge production are forgotten. Such an assessment precludes any engagement with the role of the media, the power of the state or forms of agenda-setting. It leaves no room for problematising the effects of framing that might have lasting effects on debating certain issues, and especially on the organisation of hybrid forums. For example, how a successful securitisation might shape the kinds of arguments that take place is not an issue that can be discussed by Callon and colleagues. This lack of acknowledgement of the power of framing effects is a weighty problem for researching politics.

In Chapter 5 I will show how threat narratives structure the politics of encryption. Yet, there seems to be no place for such narratives in Callon and colleagues' theory. This is especially surprising considering their critical stance towards knowledge production in traditional academic institutions. Knowledge production in hybrid forums, however, is not part of their critical analysis. Hence, the politics of setting up hybrid forums as such cannot be grasped by these authors. Brice Laurent, for example, has shown how in participatory forums 'the most vocal participants were [...] activists, who considered the national debate to be part and parcel of a science policy program they opposed' (Laurent, 2016: 778). And these activists are systematically excluded by hybrid forums since they do not conform to the image of a 'neutral' citizen. Hybrid forums presuppose neutral citizens since only they will be open to all kinds of arguments presented during the process. The creation of any public inevitably produces new forms of exclusion (Young, 2001), but these cannot be seen with Callon and colleagues' thin concept of politics. The authors are clear that hybrid forums are only ever additions to existing political institutions. Politically significant is only that which feeds back into the governmental system. Indeed, hybrid forums are political only to the extent that they feed back into the existing national political system.

Later in this chapter I will discuss how Dewey's concept of the public can be read as a more radical challenge to existing political institutions. In the next section I will discuss another approach engaging with multiple publics that employs a broader concept of politics.

The problem of issue formation

One way to open up the concept of the political has been proposed by Noortje Marres, who investigates how material objects deliver a certain kind of participation. She criticises Callon and colleagues' prescriptive theory of democracy by arguing that their commitment to Actor-Network Theory (ANT) precludes them from asking more normative questions (Marres, 2007: 763, 764). Marres accuses them of building on traditional theories of democracy when insisting that they 'commit themselves to a republican conception of democracy: they adopt a sociologized and ontologized notion of the common good' (Marres, 2007: 764). For her, the problem lies in the fact that Callon and colleagues (among others) never spell out their underlying assumptions; therefore, they are unable to reflect on their analytical vocabulary (see also the discussion in Brown, 2015). The aim of Marres's work is to expand the question of democracy and technology beyond previous liberal approaches, which are mostly concerned either with how to control science and technology in a democratic way or how to increase participation when deciding highly complex issues (Marres, 2007: 765; Marres, 2012: 52–55). She criticises previous STS approaches for uncritically rehashing the vocabulary of liberal democracy and thereby constraining the political space to the parliamentary system. Criticising spatial metaphors of the public as an already constituted space, she scrutinises devices that 'implicate' people in specific issues and thus constitute a form of material participation (Marres, 2012: ch. 2, esp. 36, 50). She looks at everyday practices of environmental politics to explain how people can become entangled in specific issues. The point is less to show if certain objects are 'good' or 'bad' for increasing participation, but rather to highlight the often ambiguous role objects play in participatory practices.

In order to grasp how the public can be understood as something related to participatory practices, Marres draws on the work of Walter Lippmann and John Dewey (Marres, 2007: 765). Although Lippmann and Dewey held contradictory positions in a debate about the role of the public, Marres reads them together and attempts to highlight their shared presuppositions. According to her, they both argue that democratic engagement can thrive only with the advent of complex issues that cannot be solved by officials.[3] She insists that pragmatists state that complex problems enable democratic politics (Marres, 2005: 209). For Lippmann and Dewey, a public is a group that emerges once its members are affected by a problem. While Dewey emphasises the involvement of the public, Lippman proposes an expertocracy (Marres, 2007: 766). However, the crucial point for Marres is that '[b]oth Lippmann and Dewey characterised democratic politics as involving a particular practice of issue formation' (Marres, 2007: 766). Thus, the public is not congruent with a political community of, for instance, a nation-state.

Rather, a public comes into being only if there is an issue at stake; it can never exist in the abstract, but must be thought of as entangled with an issue. This point of view stands in contrast with that of liberals, who conceptualise the public as an already constituted sphere.

Marres's interest lies in analysing how a public emerges if people are affected by specific issues (Marres, 2012: 52–55). Here, the affinity with STS becomes clear. The focus is on how social groups are entangled and might even come into being due to a specific object of concern (Latour, 2005 builds on this idea). This is the central issue for Marres, and a promising starting point for combining pragmatism and STS. According to Marres, to be implicated in an issue is not merely to perceive a situation as problematic; rather, it is an 'ontological trouble' (Marres, 2012: 44). Routines cannot solve that problem, so collective action is needed. The public has the crucial role of articulating issues that are not addressed by existing institutions, and the question of whether an issue is a problem is itself controversial (Marres, 2007: 771). Thus, whether a certain issue will become public depends again on how it is framed (Marres, 2007: 772). Marres states that attention should be paid to the role of institutions and material objects, as well as to the context in which a specific device is used. Only then can we understand when an issue is 'publicizing' or 'de-publicizing' (Marres, 2007: 772). Thereby, it becomes possible to gain a better idea of the kinds of practices that lead to the formation of a public and those that do not. Marres's analysis goes beyond the idea that politics happens within institutions and looks at the everyday to learn about politics in the technological society.

This is a crucial way in which Marres differs from Callon and colleagues. In her empirical analyses of participation through devices, she shows how the effect of devices depends on how they are embedded in the issue (Marres, 2012: 79, 130). She is more sensitive about the processes that make an issue a public issue and more sensitive to linguistic features and context than Callon and colleagues. Her critical attitude towards STS's application of political theory raises awareness of the vocabulary used when engaging with material politics. Moreover, using Dewey and the pragmatist tradition helps to put the tradition of STS in historical perspective, which is a valuable corrective to misplaced claims to novelty. Reading Dewey also allows us to widen our perspective and include non-deliberative theories of democracy. Through her reading of Dewey, Marres enjoys better engagement with pragmatist philosophy and, importantly, democratic theory. This enables a rethinking of the notion of where to find participatory practices and thus what counts as political. However, Marres also insists on the distinctiveness of material participation. In Chapters 5, 6 and 7 I propose that this kind of distinction is not particularly helpful when discussing material politics. Indeed, Marres reads Dewey as an STS scholar *avant la lettre*. The 'distinct' character of STS must be retained. Hence, she insists on the need to 'understand material forms of engagement as a particular modality of participation, one that can be distinguished from and compared to other forms' (Marres, 2012: 1). This insistence that material participation is a distinct form of politics that can be captured only by an STS approach comes at the cost of reducing the role of politics to the one it plays in stabilising actor networks. The normative assessment of material politics is then reduced to the question of what kind of participation they support. Marres's reconstruction of Dewey focuses on his ideas on technology and the public (not always clearly distinguishing between the concepts of the public and participation) – that is, almost exclusively on the first 30 pages of *The Public and Its Problems*. [4] This reconstruction therefore excludes the possibility of reading Dewey in a more radical way – namely, as a theorist who is critical of established structures of the exercise of power. This is why it is necessary to revisit Dewey's theory of democracy and highlight its inherent critical aspects.

Marres importantly opens up a space in which the location of participatory practices can be reconsidered. Participation, as such, is not necessarily tied to state politics and deliberation within *the* public sphere. Publics can emerge in remote spaces, be mediated through all kinds of objects and assemble around all kinds of scientific and technological issues. Forms of publicness do not necessarily emerge in the context of state politics. Indeed, as the next section will show, publics can also be conceptualised as being in radical *opposition* to the state. The perspective then shifts away from the role of experts to the potential of publics to enhance democracy by questioning established political routines. This conceptualisation speaks to the overall interest of this book in political practices outside established institutions and is compatible with my discussion on the importance of everyday practices in the previous chapter.

John Dewey's theory of democracy

John Dewey's political monograph *The Public and Its Problems* (Dewey, 1999) was originally published in 1927. I believe that Marres is a little too eager to read Dewey as an STS scholar; by contrast, reading him primarily as a theorist of democracy allows us to acknowledge his ideas on politics and thereby determine how his work can be used as part of material politics. First, I will try to reconstruct Dewey's theory before returning to the initial theme of how to think about democracy in the technological society. It is thereby possible to reconstruct Dewey's ideas on democracy more thoroughly.

The distinction between the public and the private is crucial for Dewey's ideas on the public and the state. He starts his enquiry into the state and political behaviour, two concepts that are interlinked in his conceptualisation, with 'a flank movement' – namely, by looking at the nature of human behaviour (Dewey, 1999: 8). His starting point is the observation that all human actions have consequences and 'some of these consequences are perceived, and their perception leads to subsequent effort[s] to control action so as to secure some consequences and avoid others' (Dewey, 1999: 12). These consequences might affect only the persons involved in that action, or they may impact third parties. Dewey then distinguishes between public and private actions, with the former defined as those that 'affect others beyond those immediately concerned' (Dewey, 1999: 12). This distinction refers to the extent of the consequences of particular actions or, more precisely, their recognition. This point of recognition is important: the criterion for a problem being 'public' is thus not the 'objective' fact that actions might have negative consequences for third parties, but that they are *perceived* as such. A public thus comes into being if indirect consequences are recognised and the need for regulation becomes apparent:

> The public consists of all those who are affected by the indirect con-sequences of transaction to such an extent that it is deemed necessary to have those consequences systematically cared for.
>
> (Dewey, 1999: 15–16).

The public/private (*öffentlich*) distinction leads Dewey to define *a* public (*Öffentlichkeit*). Since several issues may be recognised simultaneously, it is possible for a single person to belong to several publics at the same time (Kettner, 1998: 61).

So far this sounds very much like methodological individualism, in which political action amounts to the problem of collective decision-making. How-ever, methodological individualism (and the accompanying notion of caus-ality) is one of the main mistakes a social scientist can make, according to Dewey (1975: 236; 1999: 20). For him, the individual is always embedded in social relations. Indeed, 'Dewey categorically rejects this notion of the pre-social individual' (Bernstein, 2010: 70). While Dewey opposes the idea of any mysterious 'collective agency', he insists that what individuals want and need – their preferences and identities – are only constituted in their relations towards others. Even though his analysis takes as its starting point the individual, this is not the methodological individualism of, for instance, Rational Choice Theory. It is rather motivated by a critique of concepts such as collective agency, which for him have no resonance in life; indeed, they tend to obscure the analysis of social scientists (For a critique of empiricism in IR, see: Koddenbrock, 2017; Kratochwil, 2007.)

Dewey's focus on the entanglement of humans in networks of other humans and non-humans resonates with STS (Dewey, 1999: 25, 30). However, he makes a crucial distinction between *associations of people* and *associations of objects*. He points out that people can become aware of their own role, and this self-reflexivity distinguishes them from associations of objects and animals. This reflexiveness is crucial for his conception of the public:

> When we consider the difference [between these two forms of associations] we at once come upon the fact that the consequences of conjoint action take on a new value when they are observed. For notice of the effects of connected action forces men to *reflect upon the connection itself.*
>
> (Dewey, 1999: 24; emphasis added)

This point of reflexivity might seem trivial, but it is the focal point of much of the criticism of ANT (Fuller, 2000) and indeed efforts have been made to provide a post-ANT framework that works with stronger notions of reflex-ivity (Gad and Bruun Jensen, 2010). But for Dewey a public exists not only if people are affected by something: they also need to form a group and become aware of it (Dewey, 1999: 149). Thus, a public exists only if people relate themselves to one another and to a particular issue, and if they are aware of these relations. The public does not equal the number of people affected by an outcome, nor can one 'find' it in the media. For Marres, the public is reduced to either the number of people who come together or the actions that may be observed. Since the public is theorised from an empiri-cist stance, she does not include the self-reflective aspect that in Dewey's work is so crucial to understanding the distinct character of social life.

Before elaborating on the reasons for what Dewey calls the 'eclipse of the public', it may be useful to consider his understanding of the state and democ-racy. While there are many forms in which people can associate, for Dewey the state as a highly organised public emerges if the unintended consequences of actions by other associations (businesses, families, religious groups) can no longer be regulated by those associations (Kettner, 1998: 61; Dewey, 1999: 27). The state is like a formalised public that has found itself. Special 'agencies and measures must be formed if they are to be attended to; or else some existing group must take on new functions' (Dewey, 1999: 27). If there are officials who act on behalf of those who are affected, a state comes into being. In this sense, the state fulfils the function of responding to problems, and state officials have the task of organising how publicly perceived problems are addressed. The state as a highly organised public is, however, only one side of the coin. The other, for Dewey, is the nature of the state, which, because of its high degree of organisa-tion, is unable to generate new publics. Indeed, he warns 'against identifying the community and its interests with the state or the politically organized commu-nity' (Dewey, 1999: 15; see also Dewey, 2008). A true public will always be in tension with the state because the existing institutions will always hinder the emergence of new publics. 'Inherited political agencies' will always 'prevent that development of new forms of the state' (Dewey, 1999: 31). Though the state is de facto an organised public, it should be more open to experimentation: 'By its very nature, a state is ever something to be scrutinized, investigated, searched for' (Dewey, 1999: 31).

Dewey explicitly points out that the public always emerges in *opposition* to existing political institutions. Although he conceptualises the state as the public that is sufficiently consolidated to have officials and representatives, he insists that these institutions need to be remade as soon as they are estab-lished (Dewey, 1999: 32). The existence of solidified structures is the core problem for a well-functioning democracy, for they will never be able to cap-ture new problems and new publics. Importantly, Dewey sees the reason for this in the societal change caused by technological developments (Dewey, 1999: 30). New issues emerge and demand new publics as the environment alters. This point is crucial from the perspective of material politics: techno-logical change will always require constant adaptation on the part of political and societal institutions. The spread of networked technology made encryption a crucial technology. Encryption controversies emerge in a context in which networked technology makes new monitoring practices possible, and these practices are challenged by activists, which makes encryption a public issue.

We can see that Dewey takes a rather critical stance on existing forms of democracy (for a more detailed discussion, see: Jörke, 2003). Because of technological change and the global nature of many problems, traditional institutions of liberal democracy are unable to respond to new problems. They are unable to allow new publics affected by new problems to emerge. Thus, Dewey sees a need to challenge existing institutions and explore new ways in which the public may find itself: 'The need is that the non-political forces organize themselves to transform existing political structures: that the divided and troubled publics integrate' (Dewey, 1999: 129). Dewey's critique of liberalism is pronounced and aims to challenge these traditions by facilitating experiments with new democratic forms. He sees the doctrine of liberalism and its link to private property as especially problematic (Dewey, 1999: ch. 3).[5] For Dewey, institutions such as the electoral college and even the election of representatives made sense at a specific time in history, but they might not be the only appropriate means for today's world. Abstract ideas as they are laid down in a constitution will not automatically lead to a stable democratic culture (Dewey, 1939: 34). Democratic institutions such as voting are problematic because they suggest individual freedom and autonomy while reducing democratic life to choosing one of two options (Dewey, 1999: 120). But these ideas now seem so self-evident that alternatives cannot even be imagined (Dewey, 1999: 119). Dewey argues that people still vote out of habit rather than in a genuine expression of will (Dewey, 1999: 135). He criticises the role of party politics and the fact that elections are often more about personalities than issues (Dewey, 1999: 122, 123). Thus, focusing on specific democratic institutions will weaken democracy rather than strengthen it. With regard to his own time (and I think his diagnosis would be no different today), he asserts that the public cannot find itself because most attention is paid to almost sacred institutions, such as voting or the constitution. The result is apathy on the part of the people and disregard on that of the state. Taken together, this apathy and the naturalisation of existing democratic structures preclude the development of true democracy. Dewey thus invites us to rethink the fundamental vocabulary on which our thinking about politics relies. In Chapter 7 I will take up this strand again with a more explicit focus on what it means for Security Studies when reflecting on how encryption controversies can be read as a challenge to the prevailing narrative of national security and sovereignty.

But what are the other reasons for the weak state of democracy? In 1927, Dewey believed that the complexity of issues was one of the main reasons for the parlous state of democracy. Because problems tend to be as multifaceted as their consequences are diffuse, the 'resultant public cannot identify and distinguish itself' (Dewey, 1999: 126). The reason for this complexity lies not only in the changing nature of media and technological development but also in global interconnectedness (Dewey, 1999: 127, 128). Dewey links political action to the state, but he is aware of the global nature of many problems.[6] For him, this global character is one of the main reasons for the 'eclipse of the public', alongside technological developments and a decline in face-to-face communication for organising political action. The increasing scale and interconnectedness of issues make it difficult for the public – the group of affected people – to find itself:

> An inchoate public is capable of organization only when indirect con-sequences are perceived, and when it is possible to project agencies which order their occurrence. At present many consequences are [...] suffered, but they cannot be said to be known, for they are not, by those who experience them, referred to their origins. It goes, then, without saying that agencies are not established which canalize the streams of social action and thereby regulate them. Hence the publics are amorphous and unarticulated.
>
> (Dewey, 1999: 131)

Because of this complexity, too many issues need to be solved. As a result, no public can emerge. Dewey attributes the lack of democracy to the fact that, while technologies and thus society change, ideas and symbols do not. He is not critical of technological development, as such. The crucial problem is that political institutions did not develop at the same pace (Dewey, 1999: 142). Dewey's critique of liberalism rests to a large part on the way in which liberal ideas and institutions become naturalised and unquestioned. Increasing tech-nological complexity and a multiplicity of issues that deserve attention are neither positive nor negative for democracy. Technological change poses the central problem for which democratic solutions need to be found. These solutions have to be adapted constantly.

Solidified structures will prevent the public from being able to find itself, which is why a state is seen as problematic. Because of its stable structures, it will reinforce existing structures instead of permitting the emergence of a new public. 'To form itself, the public has to break existing political forms' (Dewey, 1999: 31). Thus, finding new political forms is a never-ending experiment and exploration of new worlds. This exploration can never be a non-reflexive rou-tine (Dewey, 1999: 18). It is always about finding radical new ways in which the public can come together and act. In true pragmatist fashion, Dewey adds that how such a state can exist is not a theoretical question but something that can be demonstrated only in practice (Dewey, 1999: 32).

To conclude: according to Dewey, democracy does not flourish because traditional institutions are glorified and new ideas are not explored. Techno-logical change demands the exploration of new democratic practices. This exploration of new worlds is much more radical and essential for democracy than Callon and colleagues' hybrid forums. The aim is not merely to improve existing democratic structures or enhance the dialogue among experts and laypersons. Dewey's exploration of new worlds has the potential to present a radical challenge to existing political institutions. In contrast to Marres, I emphasise Dewey's critical stance towards established democratic institutions and his critique of the reification of specific democratic ideas that prevent the creative exploration of new worlds. His thicker concept of politics allows for an analysis that includes the power of the media and non-state actors, but also accounts for framing effects and how pre-existing social structures shape political struggles. Indeed, since Dewey invites us to go beyond the fetishisation of established practices, such as elections and opinion polls, it becomes possible to rethink what constitutes political practice.

Publicness

The work of Callon and colleagues and Marres demonstrates the value of looking at socio-technical controversies as the place where materiality and contests over values come together. Meanwhile, Dewey allows us to locate the emergence of multiple publics outside the realm of political institutions and governmental decision-making. These forms of publicness reveal their force in the form of contesting state politics, not by being linked back to governmental decision-making processes. Reading Dewey as a critical theorist highlights the way in which he criticises the focus on established democratic practices, such as elections, which become a routine rather than provoking politics. This idea of locating politics outside existing institutions, an awareness of the broader societal context and the formation of new subject positions is what I term a thick understanding of politics. Subsequent chapters will show that controversies emerge in remote places and that modes of publicness are issue-specific and involve multiple actors.

However, for this project, Dewey's ambitious notions of what constitutes a public are less helpful. They risk framing the question as the extent to which encryption controversies live up to a particular standard of 'a public'. Dewey's theory is prescriptive and, as he himself admits, does not reflect any existing democratic system. Indeed, for him, democracy is something we need to strive for, not an existing state of affairs. One way to progress from here would be to operationalise his theory and try to measure the extent to which current publics come close to these normative standards. But the aim of this study is not a normative evaluation of publicness. Rather, I am interested in investigating the spaces where controversies emerge and how we can understand them as forms of publicness. Thus, we must shift from 'multiple publics' to 'publicness', but not understood as an existing public that conforms to the standards set by Dewey. Notions of publicness will highlight the emerging character of publics, their instability and their heterogeneity. Therefore, contrary to Janet Newman (2007: 28), I do not understand 'publicness' as a normative value. Instead, I use the term to highlight the unstable and open-ended character of emerging multiple publics. In this way, the notion carves out analytical space for investigating controversies revolving around encryption. Publicness, read in the 'radical' tradition of Dewey, allows us to understand these forms of controversy as enacting what Fraser (1990) calls a 'counter-public'. I return to this issue in Chapter 7, where I engage with how encryption controversies challenge but also reaffirm hegemonic ideas about national security.

The notion of publicness speaks to research within CSS that has analysed the changing configuration of 'public' and 'private' and how security practices have helped to constitute this distinction (Walters and D'Aoust, 2015: 48). As discussed earlier, security is not only a public good provided by the state. Indeed, security practices enter the private realm, and both public and private actors provide security. This concerns not only the relative importance of privacy in relation to state actors, but also how the emergence of private actors alters perceptions of 'security' means as well as the very notions of public and private (Leander and van Munster, 2007). Because of this reconfiguration, publicness occurs in spaces such as bureaucracies or within highly technical debates (see Chapter 6). The question is thus not how these forms of publicness conform to a normative standard of democratic theory, but how it might be possible to open up an analytical space to identify emerging forms of publicness that cannot be grasped by a traditional political–theoretical vocabulary. The notion of publicness can then be used to draw on the concept of controversy by highlighting the political implications of technological controversies. Through the notion of publicness we can understand how controversies can challenge established political categories.

In subsequent chapters I will investigate technological controversies and discuss their political significance. This will allow me to show how emerging publicness reaffirms – but also challenges – established institutions.

Conclusion: controversies and new forms of publicness

Studying material politics means looking at the ways in which objects and artefacts are entangled with politics. Though not every technology or scientific discovery is understood to be political to the same degree, the political character of technology comes to the fore in socio-technical controversies. This chapter built on the work on material politics and methodological remarks made in the previous chapter by extending the debate on publics in the context of IR researching technology. To this end, I looked at STS scholars' work on the concept of controversy. First, I discussed Callon and colleagues' idea of hybrid forums. These authors propose such forums as a way in which different ways of knowledge production can be used to explore the solution of societal problems. Their approach relies on a rather thin concept of politics in which political controversies ultimately need to be channelled through state institutions. Marres's discussion of material participation is more sensitive to contextual aspects and introduced Dewey's work to STS. In line with Dewey, she understands publics as the result of issue formation, not as its precondition. Returning to Dewey's original theory opens up a way to think about not only the effects of propaganda, for instance, but also threat constructions as part of material politics. Most important for the discussion around encryption, however, is that Dewey highlights that established notions of politics can be challenged through publics. For him, it is a core idea that established institutions and doctrines should not be immune to challenge if they prevent the constitution of new publics. This radical idea of challenging political imaginaries has not yet been taken up within STS, but it is crucial for engaging with security.

Yet, Dewey has high standards regarding what is required for a 'true' public to emerge. In order to highlight that the aim of this book is not to enquire into the existence or non-existence of such a true public, I introduced the notion of publicness to emphasise the emerging character of publics, their instability and their heterogeneity. Through this concept it becomes possible to retain Dewey's critical edge in thinking about alternative political imaginaries without trying to identify a public that upholds his normative standards. Again, the idea is not to make a claim about the emergence of a completely novel form of political order. Rather, the concepts of controversy and publicness are used in a more modest way as analytical tools to discover how encryption controversies acquire political significance outside established political institutions.

In the next chapter I introduce the basic technological features of encryption and describe the core issues that must be addressed when debating it. This discussion not only lays the foundation for Chapters 5, 6 and 7 but shows how technological controversies occur both inside and outside governmental institutions.

Notes

1   This also means that I am not particularly interested in the way in which 'secrecy' prevents controversies. Rather, in Chapter 5 I will highlight how references to secrecy and unknowns play out in encryption controversies (see Aradau 2017).
2   Callon and colleagues do not elaborate on why democracy in general or the specific form they advocate is normatively desired.
3   It is unclear why Marres ascribes this view to Dewey. She notes that he holds these views but with the 'crucial modification' that Dewey 'characterised such issues in more "objective" terms' (Marres, 2007: 767). As I show below, it seems more accurate to say that Dewey's vantage point is different. For him, the increase in complex problems is merely a matter of fact, rather than something that is either 'good' or 'bad' for democracy.
4   Another example of this rather narrow reconstruction is the treatment of the state. Marres (2007, 770) takes issue with the idea that 'Dewey conceives of the state as a unitary entity'. For Marres, this assumption is not very convincing in a globalised world. According to her, Dewey overlooks the many entities that might claim authority to settle disputes. Dewey, however, was concerned with the effects of globalisation and how the emergence of various institutions complicates politics. He even published on the issue of international relations and, as I show below in more detail, his concept of the state is rather more complex than Marres's analysis suggests.
5   I am aware that Dewey can also be read as a forebear of liberal democracy. However, some of his critiques are rather stark. When discussing the history of democracy, for instance, he claims: 'So-called liberal parties were those which strove for a maximum of individualistic economic action with a minimum of social control, and

did so in the interest of those engaged in manufacturing and commerce. If this manifestation expresses the full meaning of liberalism, then liberalism has served its time and it is socially folly to try to resurrect it' (Dewey, 2008: 297)
6 The global dimension is explicitly discussed by Dewey. Thus, I am unsure why Marres accuses him of overlooking it.

Bibliography

Aradau, C. (2017). Assembling (non)knowledge: Security, law, and surveillance in a digital world. *International Political Sociology* 4(1): 327–342.

Balzacq, T. (2005). The three faces of securitization: Political agency, audience and context. *European Journal of International Relations* 11(2): 120–171.

Balzacq, T. (2015). Legitimacy and the 'logic' of security. In: Balzacq, T. (ed.) *Contesting Security: Strategies and Logics*. PRIO New Security Studies. Abingdon and New York: Routledge, pp. 1–9.

Bernstein, R.J. (2010). John Dewey's vision of radical democracy. In: *The Pragmatic Turn*, Vol. 1. Cambridge: Polity, pp. 70–88.

Best, J. and Gheciu, A. (2014a). Theorizing the public as practices: Transformations of the public in historical context. In: Best, J. and Gheciu, A. (eds) *The Return of the Public in Global Governance*. Cambridge and New York: Cambridge University Press, pp. 15–44.

Best, J. and Gheciu, A. (eds) (2014b). *The Return of the Public in Global Governance*. Cambridge and New York: Cambridge University Press.

Bigo, D., Bondiiti, P. and Olsson, C. (2010) Mapping the European field of security professionals. In: Bigo, D. *et al.* (eds) *Europe's 21st Century Challenge: Delivering Liberty*. Farnham andBurlington, VT: Ashgate, pp. 49–63.

Brem-Wilson, J. (2017). La Vía Campesina and the UN Committee on World Food Security: Affected publics and institutional dynamics in the nascent transnational public sphere. *Review of International Studies* 43(2): 302–329.

Brown, M.B. (2015). Politicizing science: Conceptions of politics in science and technology studies. *Social Studies of Science* 45(1): 3–30.

Buzan, B., Wæver, O. and de Wilde, J. (1998). *Security: A New Framework for Analysis*. Boulder, CO: Lynne Rienner.

Callon, M., Lascoumes, P. and Barthe, Y. (2009). *Acting in an Uncertain World: An Essay on Technical Democracy*. Inside Technology. Cambridge, MA: MIT Press.

Dahlberg, L. (2007). The internet, deliberative democracy, and power: Radicalizing the public sphere. *International Journal of Media and Cultural Politics* 3(1): 47–64.

Dean, J. (2001). Publicity's secret. *Political Theory* 29(5): 624–650.

Dewey, J. (1939). *Freedom and Culture*. New York: Publisher.

Dewey, J. (1975). The ethics of democracy. In: Boydston, J.A. (ed.) *The Early Works, 1882–1898*, Vol. 1: *1882–1888*. Carbondale: Southern Illinois University Press, pp. 225–249.

Dewey, J. (1999). *The Public and Its Problems*. Athens, OH: Swallow Press.

Dewey, J. (2008) Democracy is radical. In: Boydston, J.A. (ed.) *The Collected Works of John Dewey*, Vol. 11: *The Later Works, 1925–1953*. Carbondale: Southern Illinois University Press, pp. 296–299.

Elam, M. and Bertilsson, M. (2003). Consuming, engaging and confronting science: The emerging dimensions of scientific citizenship. *European Journal of Social Theory* 6(2): 233–251.

Elshtain, J.B. (1981). *Public Man, Private Woman: Women in Social and Political Thought.* Oxford: Robertson.

Fischer, F. (2000). *Citizens, Experts, and the Environment: The Politics of Local Knowledge.* Durham, NC: Duke University Press.

Fraser, N. (1990). Rethinking the public sphere: A contribution to the critique of actually existing democracy. In: Calhoun, C. (ed.) *Habermas and the Public Sphere.* Cambridge, MA, andLondon: MIT Press, pp. 109–142.

Fuller, S. (2000). Why Science Studies has never been critical of science: Some recent lessons on how to be a helpful nuisance and a harmless radical. *Philosophy of the Social Sciences* 30(1): 5–32.

Gad, C. and Bruun Jensen, C. (2010). On the consequences of post-ANT. *Science, Technology and Human Values* 35(1): 55–80.

Geuss, R. (2013). *Privatheit: Eine Genealogie.* Berlin: Suhrkamp.

Harding, S.G. (2002). *Is Science Multicultural? Postcolonialisms, Feminisms, and Epistemologies.* Race, Gender, and Science. Bloomington: Indiana University Press.

Hegemann, H. and Kahl, M. (2016). (Re-)politisierung der Sicherheit? *Zeitschrift für Internationale Beziehungen* 23(2): 6–41.

Honig, B. (2017). *Public Things: Democracy in Disrepair.* New York: Fordham University Press.

Huysmans, J. (1998). The question of the limit: Desecuritisation and the aesthetics of horror in political realism. *Millennium: Journal of International Studies* 27(3): 569–589.

Huysmans, J. (2014). *Security Unbound: Enacting Democratic Limits.* London and New York: Routledge.

Irwin, A. (2001). Constructing the scientific citizen: Science and democracy in the biosciences. *Public Understanding of Science* 10(1): 1–18.

Irwin, A. (2014). From deficit to democracy (re-visited). *Public Understanding of Science* 23(1): 71–76.

Jasanoff, S. (2003). Technologies of humility: Citizen participation in governing science. *Minerva* 41(3): 223–244.

Jasanoff, S. (2005). 'Let them eat cake': GM foods and the democratic imagination. In: Leach, M., Scoones, I. and Wynne, B. (eds) *Science and Citizens: Globalization and the Challenge of Engagement.* London: Zed Books, p. 183.

Jörke, D. (2003). *Demokratie als Erfahrung: John Dewey und die politische Philosophie der Gegenwart.* Wiesbaden: Westdeutscher Verlag.

Kettner, M. (1998). John Deweys demokratische Experimentiergesellschaft. In: Brunkhorst, H. (ed.) *Demokratischer Experimentalismus: Politik in Der Komplexen Gesellschaft.* Suhrkamp Taschenbuch Wissenschaft 1369. Frankfurt am Main: Suhrkamp, pp. 44–66.

Koddenbrock, K. (2017). Totality and practices: The dangers of empiricism and the promise of a 'logic of reconstruction' in practice- based research. In: Jonas, M. and Littig, B. (eds) *Praxeological Political Analysis.* Abingdon: Routledge, pp. 107–115.

Kratochwil, F. (2007). Of false promises and good bets: A plea for a pragmatic approach to theory building (the Tartu Lecture). *Journal of International Relations and Development* 10(1): 1–15.

Latour, B. (2005). From Realpolitik to Dingpolitik, or how to make things public. In: Latour, B., Weibel, P. and Zentrum für Kunst und Medientechnologie Karlsruhe (eds) *Making Things Public: Atmospheres of Democracy.* Cambridge, MA: MIT Press, pp. 14–41.

Laurent, B. (2016). Political experiments that matter: Ordering democracy from experimental sites. *Social Studies of Science* 46(5): 773–794.

Leach, M., Scoones, I. and Wynne, B. (eds) (2007). *Science and Citizens: Globalization and the Challenge of Engagement*. 2nd edition. London: Zed Books.

Leander, A. and van Munster, R. (2007). Private security contractors in the debate about Darfur: Reflecting and reinforcing neo-liberal governmentality. *International Relations* 21(2): 201–216.

Lezaun, J. and Soneryd, L. (2007). Consulting citizens: Technologies of elicitation and the mobility of publics. *Public Understanding of Science* 16(3): 279–297.

Marres, N. (2005). Issues spark a public into being: A key but often forgotten point of the Lippmann–Dewey debate. In: Latour, B. and Weibel, P. (eds) *Making Things Public: Atmospheres of Democracy*. Cambridge, MA: MIT Press, pp. 208–217.

Marres, N. (2007). The issues deserve more credit: Pragmatist contributions to the study of public involvement in controversy. *Social Studies of Science* 37(5): 759–780.

Marres, N. (2012). *Material Participation: Technology, the Environment and Everyday Publics*. Basingstoke and New York: Palgrave Macmillan.

Newman, J. (2007). Rethinking 'the public' in troubled times: Unsettling state, nation and the liberal public sphere. *Public Policy and Administration* 22(1): 27–47.

Rouse, J. (1996). *Engaging Science: How to Understand Its Practices Philosophically*. Ithaca, NY: Cornell University Press.

Stilgoe, J., Lock, S.J. and Wilsdon, J. (2014). Why should we promote public engagement with science? *Public Understanding of Science* 23(1): 4–15.

Walters, W. and D'Aoust, A.-M. (2015). Bringing publics into Critical Security Studies: Notes for a research strategy. *Millennium: Journal of International Studies* 44(1): 45–68.

Wynne, B. (2006). Public engagement as a means of restoring public trust in science: Hitting the notes, but missing the music? *Public Health Genomics* 9(3): 211–220.

Young, I.M. (2001). Activist challenges to deliberative democracy. *Political Theory* 29 (5): 670–690.

4 Digital encryption

Digital cryptography is not only increasingly crucial for commercial security products but also a rapidly expanding academic discipline. Although cryptography is an ancient technique, it is also a core technology in the context of digitalisation. The origins of an engagement with digital encryption as a political technology can be traced back to the early hacking culture. The term 'hacker' traditionally refers to an 'amateur tinkerer, an autodidactic who might try a dozen solutions to a problem before eking out a success' (Galloway, 2006: 151). In the early days, a typical hack might have consisted of trying to manipulate a telephone line so that calls could be made for free. Even though they are now negatively associated with criminal or even terrorist activities (Galloway, 2006: 157), hackers originally emphasised the playful aspects of their hobby as well as its political and economic implications (such as allowing them to make free phone calls). Imaginaries of an alternative political order within the hacking culture can be found to this day in '"hacktivism" – the combination of hacking techniques with political activism' (Taylor, 2005: 626; see also Coleman, 2011). Denial of service attacks, for instance, are used to draw public attention to and protest against the actions of certain actors. The issues vary and also concern 'offline' life. However, encryption is a particularly interesting topic, for here we can see the struggle between the state trying to exert control and activists' efforts to keep the worldwide web 'anarchic'. Diana Saco, in her analysis of cyber-democracy, conceptualises the political debates about encryption as a 'competition between [...] visions for restructuring cyberspace [...] and develop[ing] it in ways that make it safer for particular kinds of political agents' (Saco, 2002: 142).

It is important to understand that encryption was a subversive technology when activists first envisioned its radical potential in the early 1990s. Nowadays, however, it is used by everybody who engages with networked technology on a daily base. It is applied to secure all internet traffic (represented by the 's' in https://) and will become even more important as the internet of things continues to develop. But the pervasive nature of this technology does not make it any less contested, as will become clear throughout the rest of this book. Encryption thus always was and remains a political technology on several levels: it is an object of contention; it is a tool that is used by political actors such as journalists, state agencies and activists; and it allows for a distinct kind of (political) communication by making anonymity possible in the internet. Several countries around the world, such as India and China, heavily regulate the use of encryption, while human rights activists and journalists rely on the availability of strong encryption. Debates on the topic involve a multiplicity of actors and occur in a variety of settings, both inside and outside governmental institutions.

The controversies that I describe in this chapter and the next discuss how state agencies and activists struggle over the meaning of encryption in a security context. As I show throughout the book, these battles over encryption are connected to broader debates about surveillance and democracy. Encryption controversies always involve multiple technological as well as societal questions. The description of these battles in this chapter will illustrate this in more detail. I introduce various instances where the political character of encryption comes to the fore. Here, we can observe the important but differentiated roles of state actors, commerce and activists. Indeed, as the quotation at the opening of Chapter 1 indicates, early activists such as Timothy May could even view encryption as a tool for an alternative world order. These 'cypherpunks' were especially concerned with digital encryption and its role in the development of the internet and social order more generally. They protested against state authority, particularly when it implied controlling and regulating technology (Assange *et al.*, 2012). The debates over encryption can be understood only in the context of a wider battle about (internet) freedom (Hellegren, 2017). Encryption is merely one locus among many in which we can observe the contestation of technology and the values it transmits.

Early struggles over the orientation of the internet took place mostly in the United States. However, in this book, I will broaden the focus by looking at Germany, too, where what is now Europe's largest hacking organisation. The Chaos Computer Club (CCC) has become a key actor in internet politics. Individual members express their views in newspapers and are regularly invited to talk in front of parliamentary committees. Since its early days, the CCC has combined hacking with more outward-oriented activities in order to inform the public about technology and its possible insecurities (Kubitschko, 2018: 90–92). One of its most famous hacks targeted the so-called *Staatstrojaner* (State Trojan) – a program that was designed to monitor communication – but the CCC proved not only that the software could be used by third parties but also that it enabled the collection of more data than is constitutionally permitted. These findings were published in a major German newspaper and show how the CCC works in tandem with established media outlets to increase its legitimacy as a civil society actor (Chaos Computer Club, 2011; Kubitschko, 2015). Hackers often shift between highly specialised activities and those that try to capture the public's attention. Organisations such as the Electronic Frontier Foundation (EFF), based in the United States, are crucial participants in the struggle over the meaning of encryption. Their efforts are not limited to suing state agencies; they also try to publicise issues related to internet politics. This theme of political action outside established institutions runs through this book. In this chapter, it will be introduced by highlighting the multiple sites in which controversies over encryption first emerged and continue to emerge.

The chapter thus fulfils two functions. First, it explains some basic technological principles to support a better understanding of the technology at hand and provides an overview of the main issues to consider when talking about the politics of encryption. For example, with no more than a rough idea of the technological underpinnings, it is possible to understand that security in the context of encryption is inevitably provisional. I provide a brief summary of the political battles and discursive tropes as a background for the following empirical chapters that focus on more specific issues. Second, this overview of the politics of encryption anticipates the conceptual arguments of subsequent chapters. It shows how political practices occur outside established political institutions, and how a framework of national security does not really allow us to analyse encryption. Not only does the technology transcend national boundaries, but standards set in the United States are now de facto global and thus affect the politics of encryption throughout the world. In sum, tracing the different controversies as well introducing sites where encryption plays a role today shows the involvement of a multiplicity of actors, loci and technological issues.

Some basic principles of encryption

In the simplest sense, encryption refers to the scrambling of letters in order to conceal information. Narrowly understood, it is a tool that allows for secrecy. However, this definition does not acknowledge the performative effects of encryption and how it is even used as an artistic tool (DuPont, 2014: 4). In the context of this book, the focus is on what encryption does for security, which is why the focus is on encryption as part of a security infrastructure in the digital age. Of course, there are other techniques for hiding the content of a message. For instance, while milk ink is a traditional way to conceal a complete text, in a digital environment anonymisation tools can be put to work to conceal the origin of a message. In practice digital encryption technology is often used in combination with anonymisation tools. Encryption only makes it possible to disguise the *content* of a message, so other tools are necessary to hide the *origin* of that message or even the fact that it has been sent. But encryption cannot only prevent eavesdropping by unwanted third parties; it can also reassure the recipient of a message that the sender is who they claim to be and that the message has not been altered. This can be achieved only by encryption programs that rely on mathematical principles – namely (as yet) unsolved mathematical problems. I will explain the basic principles on which encryption relies without going into too much technical detail (for an overview, see: Kahn, 1997; Schneier, 1996; Singh, 2000).

Encryption aims to make the original message unreadable. Hence, in order to read the message, a specific algorithm that allows the intended recipient to reverse the encryption is needed. An algorithm is a fixed sequence of steps that is taken to solve a particular set of problems. Here, it means the steps that are taken to scramble data in a particular way to make it unintelligible to unauthorised third parties but decipherable for the legitimate receiver. The classic and most simple example is the Caesar algorithm. Every letter is exchanged with another letter in the alphabet. In its most straightforward form, A becomes B, B becomes C ... and Z becomes A. 'Hello' (the so-called plaintext) would therefore appear as 'IFMMP' (the ciphertext). There are more sophisticated ways of scrambling the letters in ways that are not so easy to guess. In this example the number of keys is also limited to 26 – the number of letters in the alphabet. The text can thus be relatively easily deciphered merely by trying every possible key. Such simple algorithms can be applied manually but, given enough time, also cracked rather easily.

The likes of the Caesar algorithm rely on the fact that both parties know how to decrypt the message. That is, the receiver needs to know what to do with 'IFMMP'. The means to unscramble the message is called the key (A = B). Thus, encryption traditionally works only if both parties are in possession of the same (correct) key. In practice, this scrambling of letters – or 0s and 1s – is done in a more complex way that obscures the relationship between the ciphertext and the plaintext, but the underlying principle remains the same.

In the 1970s IBM developed DES – the Data Encryption Standard (Oppliger, 2005: 236). Today, DES is considered to be insecure since the key is too short; several tests have shown that DES-encrypted texts can be deciphered relatively quickly using an exhaustive key search (Oppliger, 2005: 250). Additional security could be gained by repeating DES three times in a single message. However, this '3DES' method was deemed too slow and was still insufficiently secure. A stronger algorithm is AES (Advanced Encryption Standard), which was the outcome of a public competition to establish the new encryption standard.[1] It was standardised by the National Institute of Standards and Technology (NIST) in 2000 and remains in use to this day (although it is now mainly used in conjunction with other programs). The procedure of determining the next most secure algorithm was public and transparent. This standard was established in the United States but it is now the de facto global standard. Hence, in the context of encryption, we can already note that states are unable to exercise complete sovereignty.[2] Standards are set on a global level and can scarcely be unilaterally changed. For instance, AES is used in the SSL/TLS protocol that one encounters whenever connecting to the internet (Dooley, 2013: 84, 94).

Symmetric encryption systems systems rely on the fact that both parties possess the key – which, ideally, they will have exchanged in a face-to-face meeting. Nowadays, however, a lot of information is exchanged between parties who will never meet in person, so the physical exchange of keys is impractical (Singh, 2000: 251). This is why 'asymmetrical' encryption has become crucial. *Symmetrical* encryption systems rely on the same key for encryption and decryption. In the previous example, the key would be 1, since each letter would be substituted by the next letter. This key would mean that each letter is replaced by the next letter in the alphabet (A = B) during encryption, and that each letter is replaced by the preceding letter (B = A) during decryption. By contrast, a*symmetrical* systems utilise two different keys during encryption and decryption – a public key for encryption and a private key for decryption. This is made possible by the fact that certain mathematical functions can be easily computed but are very difficult to reverse.

A simply metaphor helps to explain the basic principle on which public-key cryptography relies. Say Alice wants to send Bob a message that Eve should be unable to read. Alice puts the message in a box and locks the padlock with the only key. She then sends the padlocked box – but not the key – to Bob. He puts a second padlock on the box and returns it to Alice. So, the box that contains the message is now locked with two padlocks. Since Alice's padlock is still on the box, she knows that no one could have opened it and altered the contents. She then unlocks her padlock and sends the box back to Bob. Now he knows that the contents have not been altered because his padlock is still in place and he has the only key. He unlocks the padlock and reads the message.

The idea of an asymmetrical key was first published 1975, and two years later the cipher (called RSA) that put the idea into practice was developed (Salomon, 2006: 273, 274). Today, asymmetrical cryptography is a core feature of internet security. However, as asymmetrical keys need more computational time, most systems utilise a combination of symmetrical and asymmetrical encryption. Public-key offers better security, but a system is still needed to guarantee the identity of the person using a specific key. Hence, a so-called public-key infrastructure (PKI) is necessary. Such an infrastructure relies on third parties that verify the identity of actors. These are called Certification Authorities (CA) and their standardisation makes it possible for digital signatures to be legally binding (Kościelny *et al.*, 2013). It is important to bear in mind this idea of a third party possessing the key when contemplating the political battles around encryption.

One of the most important aspects of encryption as a security technology is that there are very few standardised encryption systems, such as AES and RSA. Some of these are used for symmetrical and some for asymmetrical encryption, while others are mostly used for signatures. Most security issues arise from flaws in *application* as opposed to problems with the underlying algorithm itself. Thus, most reports of flaws in encryption refer to a specific application, such as SSL. Establishing the security of these applications is a multifaceted process. In any case, however, the basic system relies on mathematical principles. Security is the function of mathematical problems to which no solutions have yet been found. For example, it is quite easy to multiply two prime numbers, but reversing that procedure is much more time-consuming. If one were able to find a quick and easy way to do so, then all of the encryption that relies on that particular problem would become insecure. This implies that the security of encryption can never be positively proved. Indeed, the only 'proof' of any algorithm's security is that many people have tried and failed to break it. This is another reason why establishing a new standard is so difficult: tried and tested systems are considered to be more secure than new ones. Encryption security is thus never fully established, but always rests on uncertainty: the technology only ever allows for secure communication until proven otherwise. This idea of incomplete security and uncertainty will be discussed further in Chapter 5, where I will show how themes of uncertainty play out in the discourse on encryption. This point also shows that debates about encryption and security inevitably need to acknowledge technological features. Contesting encryption and its role in surveillance and security practice also means contesting its technological aspects. However, before describing these controversies in more detail, a brief history of encryption is in order.

From war technology to blockchains

Although encryption is an ancient technology, it was only the invention of the telegraph, which permitted rapid communication, that led to its widespread use in the nineteenth and twentieth centuries (Dooley, 2013: 31; see also Singh, 2000). Especially during the American Civil War, new kinds of cryptography and the means to decipher them (cryptanalysis) were developed. Encryption was always tightly linked to security concerns as it was mainly perceived as a war technology. Deciphering encryption at this time was done with pen and paper and mostly relied on knowing the relative distribution of letters in a specific language (e.g. 'e' is the most common letter in English, and certain combinations of letters – such as 'ie' and 'th' – occur much more often than others). The development and dissemination of radio enabled another mode of analysis that continues to be used to this day – traffic analysis (Dooley, 2013: 50–51). This looks at where a message originated and to whom it was sent. Such information can be rather telling as current debates about metadata, which includes details of who communicated with whom and when, show.

In World War II, all of the major combatants used simple 'computers' – or, more precisely, rotor machines – for encryption, of which the German 'Enigma' machine is probably the most famous. The very sophisticated Enigma system relied on a combination of rotors and plugboards that complicated deciphering. However, Alan Turing and the rest of the codebreaking team at Bletchley Park were eventually able to decipher the code (Dooley, 2013: 7). It was only with the development of the internet almost half a century later that encryption came to be used for civilian purposes on a large scale. It was implemented in digital technologies as e-commerce emerged and quickly became a relevant factor in the economy more generally and especially in encryption debates. With the spread of digital communication, the problem of intervention by third parties became ever more important. Today, everybody who uses networked technology inevitably uses at least basic encryption, in which information is scrambled to make it inaccessible to third parties. However, greater computational power has facilitated the development of more complex algorithms.

With the increased use of the internet and especially e-commerce, it became necessary to have a standard of encryption to allow for interoperability and avoid fraud. The first of these was DES, which was widely used, although from its early days there were rumours that the National Security Agency (NSA) influenced its design. It is now widely acknowledged that the size of the key was reduced at the NSA's instigation, which weakened DES. There is less certainty about the NSA's influence on another part of the algorithm. Roughly speaking, the so-called S-boxes define the way in which the data is scrambled. These are flawed in DES, but the extent to which the NSA exerted influence over this part of the design is not known. Although DES was long considered to be insecure, US government agencies continued to insist that it was secure (EFF, 2016). However, ultimately it was proved to be too weak, since its key length is too short and prone to brute force attacks. An attacker simply has to try every key until they find the right one. The contestation over the security of DES is an early example of the involvement of state government and civil actors in debates over encryption. Indeed, the whole dispute revolved around whether a certain technology could be considered secure. Note that the question of what constitutes security is addressed through the question of assessing a particular technology. This is the core theme of Chapter 6, but here we can already see how societal and technological issues are interwoven.

Indeed, the debate over the security of DES was linked to the broader battle over who should control cryptography. Since it was considered to be a war technology, it was argued that it should remain under state control, so policies aimed at strengthening control over encryption were implemented throughout the 1980s (Dunn Cavelty, 2007: 50, 51). However, commercial and private actors increasingly challenged these claims, which ultimately led to the crypto-wars of the 1990s (see below). The field of cryptology is becoming ever more sophisticated. Rather than being a niche, specifically military field, developing strong encryption (and breaking it) is now a major endeavour within government agencies as well as part of the rapidly increasing business of security and hacking technology (Marquis-Boire, 2013).

Public-key cryptography has developed quickly over the past decades, and more sophisticated systems, such as elliptic-curve cryptography, have emerged alongside it. Especially when we think about the rapidly expanding internet of things (i.e. devices that are networked and can be controlled remotely), encryption becomes ever more important if electronic devices are to be used in the likes of healthcare and self-driving vehicles. This also shows that the development of encryption cannot be uncoupled from the role that private actors play; nor can the politics of encryption be considered in isolation from the complex roles that commercial actors play. For example, private security companies market tools for monitoring targets, but also for breaking encryption applications. States often work with these private companies (Parsons, 2016), which offer a variety of tools that may be used to hack or circumvent encryption in the name of higher security (FinFisher, 2018; GammaGroup, 2018). I will discuss the role of commercial actors in detail in Chapter 5, mainly in the context of ICT companies cooperating with state surveillance (Lyon, 2014). Here, suffice to say that public–private partnerships – as well as less visible joint ventures between state actors and private companies – are becoming ever more crucial for governing internet security (Carr, 2016). Encryption technology must be seen in connection with the increasing commercialisation of internet security. It has developed from a niche military technology to become a cornerstone of a rapidly growing industry. The commercial role of encryption technology is a theme that will recur in subsequent chapters, and indeed it is often decisive for politics. In any case, it is crucial to remember that controversies over encryption are not fought only between states and internet activists.

Encryption is also crucial for an emerging technology that has attracted a great deal of attention in recent years: blockchains. Though not strictly related to the main theme of this book, a few remarks on this development are necessary. Blockchain technologies, all of which use public-key cryptography, enable large groups of people to implement new forms of decentralised, coordinated action. Decisions can be made only by a majority, not a single agent. Equally important is that no transaction can be altered (De Filippi and Wright, 2018: 44), and all transactions are visible to all participants (in a peer-to-peer network) and time-stamped. As a result, participants in the network do not need to know each other or even trust each other, because, in theory, tampering with the blockchain is impossible.

Financial transactions, voting and other forms of decision-making are thus possible in a highly decentralised manner (Atzori, 2015: 2), and blockchain technologies might even change the nature of legal contracts (De Filippi and Wright, 2018: 4). They are used as part of the capitalist economy, but are also seen as a way to implement alternative forms of social organisation: 'Inspired by models of open source collaboration, decentralized organizations connect people together through blockchain-based protocols and code-based systems, focusing on achieving a shared social or economic mission' (De Filippi and Wright, 2018: 136). The ideological roots of early hacking culture are evident here. Within the cryptocurrency community, actors disagree on whether to work within or outside and against capitalist structures. In this context Quinn DuPont explains that some actors envision blockchain technology as an enabler of a new form of decentralised autonomous organisation (DAO), which in turn would enable a new form of sociality that 'would be transparent, efficient, fair and democratic' (DuPont, 2017: 157).

Blockchain technologies are most prominently discussed and were originally applied in the context of cryptocurrencies. Many online retailers now accept Bitcoin as a currency, and some banks have even started issuing their own cryptocurrencies (Campbell-Verduyn, 2017: 2). Popular interest in cryptocurrencies seems to have peaked, and while much literature is narrowly focused on their technical aspects, their social impact requires more careful study (Campbell-Verduyn, 2017). Blockchain technology certainly has the potential to impact forms of organisation and decision-making processes, as well as the economy at large. Once again, this subject highlights the need for political scientists to pay closer attention to new technologies.

In summary, this section has shown that encryption grew from a military technology to one that is now ubiquitous and deeply embedded in the political economy of security. Indeed, controversies about encryption involve a multiplicity of actors, with commercial actors playing an especially decisive role. The next section will discuss the controversies over encryption in more detail to highlight not only the technological but also the social aspects of these ongoing debates.

Privacy, anonymity, freedom: encryption as a political battlefield

Cryptowars I

The term 'cryptowars' refers to a series of controversies during the late 1980s and 1990s in the United States. At the core of the debate around encryption lies the question of the extent to which law enforcement should be able to wiretap citizens' communications, or if this should be considered an illegitimate violation of privacy. My reconstruction of the role of the various actors shows how these controversies, although located in the United States, have had a global impact on surveillance and privacy practices.

The term 'wiretapping' originated as a reference to the interception of communications transmitted via copper wire, but nowadays it is used more broadly to refer to other kinds of communication, and most importantly digital communication (Diffie and Landau, 1998: 151). When copper wires were replaced by fibre optics, law enforcement agencies initially had trouble adapting to the change, since 'wiretapping' became more difficult. However, changes to legislation were quickly made and, since then, and especially with the increasing use of digital technologies in everyday life, the possibilities for surveillance have multiplied (Swire and Ahmad, 2012: 421–422). In 1978 the Foreign Intelligence Surveillance Act (FISA) was passed into law. Legally, a special FISA court needs to approve every request for wiretapping. However, these rulings have been much criticised because the court almost always approves the request (Diffie and Landau, 1998: 179, 180). Especially after the 9/11 terrorist attacks, the power of FISA expanded. Although it is supposedly only concerned with foreign intelligence, networked technology has blurred the distinction between domestic and international communications. FISA thus relies on different regimes of protection for US and foreign citizens, a distinction that is difficult to sustain in the context of globalised, networked technology. In practice, US citizens are also impacted by this surveillance regime. Furthermore, the scope of foreign intelligence increasingly includes intelligence gathering for criminal cases (Goitein and Patel, 2015). Of most concern is the secrecy under which FISA courts rule. Only one party – the government – is ever in front of the court, so it is unsurprising that the court rarely refuses a request (FISA, 2015).

With the growth of the internet, encryption became one of the few possibilities to prevent wiretapping or at least to make it more time-consuming and costly. Because encryption programs became strong tools against wiretapping, they quickly became the focal point of a public debate, the outcome of which would be decisive for the future development of the internet (Saco, 2002: 5). Today almost no one considers a complete government ban on encryption to be feasible. Encrypting communication is important for all economic transactions through the internet, and government agencies rely on strong encryption, too. Hence, the political question is: who decides who is permitted to use what kinds of encryption? Some governments tried to introduce strong regulation of encryption. However, attempts by the US government in the early 1990s to regulate the application of cryptography elicited furious responses not only from civil rights groups but also from various companies with a financial interest in the free availability of encryption. These debates are summarised under the term 'cryptowars', starting with the development of the encryption software Pretty Good Privacy (PGP) and its global distribution. Although the cryptowars were characterised by intense public debate, the NSA has been an important background actor on cryptology at least since the 1960s (Kahn, 1997, ch. 19).

PGP, which is widely used for encrypting emails, was initially considered illegal since its programmer, Phil Zimmermann, had applied patented algorithms of RSA Data Security without a licence. He released the program on the internet in 1991 and it naturally spread to other countries, although US law 'classifies cryptographic software as "munitions" and forbids its export' (Levy, 1994). Cryptography was always considered vital to national security concerns, especially in the United States, which feared that its enemies – i.e. other states, criminals and terrorists – might use strong encryption that could not be decrypted by the national intelligence services. Even at the time, Zimmermann claimed that he was afraid of an 'Orwellian future' in which the government would be able to eavesdrop on all communications (Levy, 1994). Encryption is thus a dual technology that can be used to enhance but also resist state power (and nowadays, more importantly, the power of corporations). Because PGP spread so quickly via the internet, it could not be banned by the United States. Here we see how the legal structures applied by nation-states are rendered obsolete and indeed dissolved by technology. Its dispersed character makes a national framework of direct regulation impossible.

In 1993 a lawsuit was filed against Zimmermann, but it was dropped three years later (Schwenk, 2010: 31).[3] Zimmermann adjusted the code so that it would be compatible with the free licences that RSA Data Security had just released. PGP could then be distributed legally (albeit in different domestic and international versions) and for free. To this day, PGP remains a cornerstone of privacy measures for digital communication between individuals. In 1996, restrictions on encryption were loosened, mostly due to pressure from the business community. The United States failed to promote its stance on cryptography worldwide, and German companies, for example, could increase their sales of cryptographic tools because of US restrictions (Diffie and Landau, 1998: 221). Even a report by the National Research Council found that the United States 'would be better off with widespread use of cryptography than without it' (Diffie and Landau, 1998: 220; see also Zangrilli, 2012). In the end, appeals to national security were unable to trump economic interests and the concerns of civil society. As it became clear that export controls on code were impossible to enforce in a digital environment, regulations were finally loosened. The new rules 'did not fully satisfy civil-libertarians, however, as the US government retains the ability to review the export of any software and classify offending code as a munition and forbid its export' (Herrera, 2002: 114). Moreover, encryption remains banned or at least strongly regulated in a number of countries. The law in India is currently under revision, but at present it does not permit any uncontrolled encryption stronger than 40-bit (i.e. very weak). Similar restrictions apply in China and Saudi Arabia (see Acharya, 2017; Usher, 2013). And after the attacks on the offices of *Charlie Hebdo* in Paris, British Prime Minister David Cameron entertained the idea of banning encryption in the United Kingdom (UNHCR, 2015; Yadron, 2015). Although this suggestion was quickly and widely dismissed, the idea of regulating or even prohibiting encryption persists in many countries.

In the second phase of the cryptowars, the debate shifted away from export controls to more immediate questions of privacy and surveillance. The second issue under debate was the legitimacy of implementing so-called escrow systems, which would have enabled the US government to hold a 'master key' to access all communication. This master key would, however, be split and held by two different institutions. A court order would be needed to grant access to both halves of the key, which would then enable decryption of the information. The first 'Escrowed Encryption Standard' was limited to information transmitted via telephone systems, but this was later widened to include all 'telecommunications systems'. Indeed, proposals went as far as 'Clipper phones', which had a built-in chip that would enable the government to decrypt all communication (after the appropriate court order had been obtained). In response to criticism, the government proposed that the keys would be stored not with the government itself but with a third party in order to prevent illegitimate government access. But even this precaution did not satisfy privacy advocates. For them, the underlying logic was problematic: namely, that the whole security system was based on the assumption that private users should not be trusted with their own keys (Saco, 2002: 158–159). The scheme failed: only a few organisations bought the Clipper phones, mostly the FBI, while AT&T products that incorporated the chip attained disappointing sales (Diffie and Landau, 1998: 210–222). Nevertheless, a fierce debate revolved around the question of whether an escrow system should be compulsory and whether such systems would actually aid law enforcement. Finally, in 1999, the US administration changed its position. It is usually assumed that this was due to an acknowledgement that strong encryption was needed for e-commerce, the fact that other countries would profit from the sale of cryptographic software while US companies would be excluded from this market, and the difficulty of enforcing some of the regulations. Again, commercial interests are crucial to understanding the politics of encryption. The debates about the technology are often more diffuse, involving as they do a multiplicity of actors holding varying beliefs about encryption as a technology and its political implications.

Opponents of the Clipper chip and advocates of widespread strong encryption continue to enjoy support from civil society organisations, among which the EFF remains the most prominent. Furthermore, individual hackers have been very active in opposing government policies; and the same could be said of software companies who are afraid that weak encryption will harm their position in the world market (Saco, 2002: 147). In the 1990s the (utopian) idea of the internet as a free space unregulated by the state was still at the forefront, although it eventually lost out to the commercialisation of the medium. Yet this very commercialisation made it necessary to allow for better encryption. Hackers have had a strong presence in both traditional media and politics, and in part it was this – as well as their alliance with a number of powerful companies – that enabled them to 'win' the battle over encryption. If they want to make an issue more visible to a wider public, hackers still need to engage directly with the offline world, attract traditional media attention and communicate in a way that ensures they are understood by the average user. An example of this was the 'DES Challenge', held in 1997, in which a number of small groups from around the world competed with one another to show the insecurities of DES. The members of one of these groups, DESCHALL (Curtin, 2005), organised themselves via email lists and exposed the inherent vulnerability of DES, not least because they were able to recruit so many people who were willing to run the necessary key-search on their computers. Even though thr group's target audience was limited to people who were interested in digital technologies, the activists still needed to communicate specialist knowledge to a wider audience (Curtin, 2005: 77–80). In order to mobilise more participants (and their PCs), they had to learn how to 'dumb down' the technical details. In order to convince the public of the relevance of their concerns, these activists had to bridge the gap between 'experts' and 'laypersons'. The technologisation of security thus occurs not only in security agencies or among professionals, but also in more remote spaces. In Chapter 5 I will continue to explore this topic when discussing how contemporary activists try to raise public awareness of surveillance and contest the state narrative of the need for more monitoring and demands for weaker encryption.

One of the few 'civilian' – i.e. non-governmental – advocates of the Clipper chip and other escrow systems was Dorothy E. Denning, a computer science professor who argued in favour of governmental control over cryptographic systems. While she supported authentication and digital signatures, the prospect of strong encryption that would guarantee anonymity posed, in her view, a significant threat to society (Denning, 2001b: 88). She felt that it would impede law enforcement, while terrorists, organised crime and distributors of child pornography would be able to use unbreakable encryption to remain anonymous (Denning, 2001b: 97). These dangers are still invoked in opposition to encryption, although in 2001 even Denning had to admit that in most cases encryption did little to impede law enforcement (Denning, 2001a). Nevertheless, most of the supporters of strict legislation against encryption are still found in the police and security services, with a few academics and politicians siding with them.

It is striking that even the positions of the state actors – such as the FBI, the NSA and the federal government – cannot be easily summarised. Indeed, at present, their opinions on how to govern encryption differ greatly. For example, Michael Hayden, a former director of the NSA, announced his opposition to the FBI's request to obtain the means to decrypt an iPhone (Page, 2016). While the FBI is in favour of extremely weak encryption, the NSA seems to be confident that its capabilities are good enough to circumvent even strong encryption. Hence, it openly supports the use of encryption and advocates for certain systems that it considers to be sufficiently strong. Similarly, TOR – an anonymisation tool that I will describe in more detail below – was a target of numerous attempts to weaken it, although it was and continues to be supported by the US government (TOR Project, 2016). Law enforcement and other state agencies rely on TOR as a tool for secure communication. It is thus a technology that was developed and is used by the state, but it is also utilised by dissidents to avoid state surveillance.

Since the 1990s activists have been expressing concerns about the NSA and its ability to break certain encryption systems (Levy, 1994). The fear is that government agencies have (secret) knowledge about these systems. In addition, there have always been rumours that US state actors have compromised some systems, lobbied for weak systems in the knowledge that a high level of security could not be reached (Curtin, 2005), or implemented so-called backdoors – built-in ways to circumvent encryption. Indeed, the US government has tried to implement backdoors, against opposition from civil society activists. Knowledge of these backdoors makes the decryption of encrypted data possible, which is why privacy advocates view them as problematic. If companies or states have access to them, the encryption system is compromised. As discussed in Chapter 2, we can see that the political question of what is known and what is not known becomes a major part of the controversy (Aradau, 2017; Barry, 2012; Walters, 2014).

Multiple controversies emerged during the cryptowars, and subsequent chapters will show that many of the themes that were prominent in the 1990s are still present today. Among the contested issues were the influence of encryption and what constitutes secure knowledge, but also the question of stronger regulation. The first cryptowars ended with the deregulation of encryption after plans to implement escrow systems failed. For a while, other security issues – primarily the 'War on Terror' – overshadowed the debate over encryption (see Dunn Cavelty, 2007). However, it re-emerged rapidly in the wake of Edward Snowden's revelations about the surveillance activities of the NSA and the 'Five Eyes' intelligence pact. This marked the beginning of Cryptowars II.

Cryptowars II

After the deregulation of encryption, cryptographic tools were not widely used, shattering any hope on the part of activists for a more secure – that is, a more anonymous – internet (Diffie and Landau, 2007: 257). It seemed that users did not care too much about their own privacy or security, and encryption was not really on the public agenda for several years. The political climate shifted after 9/11, and especially in the US context monitoring foreign nationals became a main security goal. The Bush administration authorised 'warrantless wiretaps between targets abroad and those with whom they communicate inside the United States' (Diffie and Landau, 2007: 302). This paved the way for massive domestic surveillance programs. Although these activities were somewhat known and debated, only Edward Snowden's revelations returned the question of privacy to the top of the political agenda. In this context of post-Snowden debates on privacy and, by extension, encryption, the notion of a 'Crypto War 2.0' was coined (Meinrath and Vitka, 2014). On the other hand, some have argued that the debate on encryption never really ceased, so it makes little sense to speak of a second set of cryptowars (Schoen, 2010; Burghardt, 2015). Either way, it is striking that many of the patterns of argumentation familiar from the 1990s can be observed again today. For instance, the idea that the widespread application of encryption will only help criminals has featured prominently in both phases of the debate. The argument that strong encryption facilitates terrorist attacks intensified after the attacks on the French satirical magazine *Charlie Hebdo* in January 2015; in San Bernardino, California, later that year; and after the attacks in Brussels in 2016. However, while some patterns of argumentation developed more than two decades ago, the discursive patterns have undoubtedly shifted. Encryption debates in the 1990s were embedded in a discourse on hacking culture with the aim of criminalising the whole movement (Hellegren, 2017). In the current debate, the security threat is imagined to come from terrorists, sex offenders, organised crime and drug dealers – and thus to have multiplied (Schulze, 2017).

As I will show in more detail in Chapter 5, encryption has also acquired a reputation for providing higher security by preventing state surveillance, and escrow systems have reappeared in political debates. As a result, both sides find it difficult to prove when encryption impedes law enforcement and when other methods of data gathering might be sufficient (Zittrain *et al.*, 2016: 11). By now, we know that the FBI exaggerated statistics on how often encryption prevented access to data (Barret, 2018; Crocker, 2018; Lewis *et al.*, 2017). Nevertheless, the argument that strong encryption obstructs law enforcement is far from dead.

The FBI again featured as an important actor during the 2016 controversy with Apple. The FBI asked Apple to decrypt a phone that belonged to one of the perpetrators of the San Bernardino attack. It was claimed that the phone's data would provide a better understanding of the context of the attack and help identify possible accomplices. Apple's opposition to the request may be seen as a PR stunt to reassure its customers of the importance the company accords to privacy.[4] An open letter to customers published after it refused to cooperate begins as follows:

> The United States government has demanded that Apple take an unprecedented step which threatens the security of our customers. We oppose this order, which has implications far beyond the legal case at hand. This moment calls for public discussion, and we want our customers and people around the country to understand what is at stake.
>
> (Apple Inc., 2016)

It does not really matter if Apple was motivated by a genuine concern for its customers' privacy or simply sensed an opportunity to boost its public image. What is important is that the company sought to make a technological issue – the effects of encryption – a matter of public concern. And it succeeded, because the case was hotly debated among the US public. Although the issue of terrorist attack is a securitised one, this has proved insufficient to prevent contestation. Apple actively tried to counter the narrative of encryption as a security threat by focusing on the impact such measures are likely to have on democracy. What is especially interesting here is that an *economic* actor contested the security logic of the FBI. This counters the assumption that economic actors depoliticise issues by imposing a technocratic logic – a topic to which I will return in Chapter 6. However, the role of business has always been ambiguous in the case of encryption. This, of course, does not negate the fact that ultimately every company is mainly interested in increasing its profits. But the position of business cannot be reduced to a single position or interest. For instance, unencrypted data is crucial for Google's business model as it allows the company to target advertisements. By contrast, Apple decided to fight the FBI because it is in its economic interests to offer strong privacy to its customers – at least once they started to care about the subject. The role of the user has thus become ever more important, and companies understand the need to fight to gain legitimacy in the eyes of those users.

I will return to this topic in Chapter 5, where I will focus on how encryption appears in surveillance debates. It emerges that encryption and security are increasingly framed as issues that the user must consider. Thus, it seems that the cryptowars are far from over. Old arguments about escrow systems are still espoused in public debates, and the establishment of state control over the development of encryption remains as controversial as ever (Cerulus, 2017; Singh Guliani, 2016).

The 'darknet', anonymity and TOR

Though it can cipher messages, encryption itself does not provide anonymity. Therefore, technologically speaking, encryption and anonymity are two separate issues. Nevertheless, they are often entangled. This will reappear in later chapters when discussing privacy and how it is played out against surveillance in encryption controversies. Thus, a few remarks on anonymity and the 'darknet' are in order to situate the political debates.

Nowadays, full anonymity when browsing the internet is difficult to achieve. Cookies and add-ons trace everybody's movements both online and increasingly 'offline' through mobile apps. The difficulty of maintaining anonymity – as well as the political contentiousness of the notion of complete anonymity – came to the fore in the aftermath of Edward Snowden's revelations. For example, Snowden had an email account with Lavabit, a company whose business model was based on offering total privacy. It eventually opted to shut down rather than comply with the demands of the US government (Lavabit, 2016). Since Snowden had used Lavabit, the authorities demanded the key that would allow access to his communications. Ostensibly because of the sensitive nature of the case, Lavabit's owner, Ladar Levison, was prohibited from discussing the matter in public. Hence, Levison found it difficult to secure the services of a lawyer who would understand the technicalities and represent him appropriately. More importantly, legal restrictions meant the case received little public attention. The court ruled against Levison, and the ban on him discussing what had happened remained in place even after he shut down Lavabit. Instead, he made the somewhat cryptic declaration: 'I have been forced to make a difficult decision: to become complicit in crimes against the American people or walk away from nearly ten years of hard work by shutting down Lavabit' (cited in Poulsen, 2013).

One of the very few tools that support strong anonymity is The Onion Router (TOR). This directs traffic through a number of relay servers that can be envisaged as the layers of an onion: the servers know only the previous location, not the origin or the final location (TOR Project, 2018). The relay servers are hosted around the world in a decentralised way – an essential feature of the system that ensures anonymity. TOR is used by human rights workers and journalists, law enforcement and criminals, but also by people who are simply interested in higher levels of anonymity.[5] Using TOR is not particularly difficult: one needs only to install the browser bundle and follow a few security rules, such as not using torrent or any browser add-ons. Importantly, the more people use TOR and the more relays there are, the more secure the system becomes. But TOR is also discussed in more negative contexts, most notably in relation to the darknet. For instance, TOR is the largest provider of otherwise inaccessible web pages. I consider 'darknet' a problematic term, since it suggests that a particular technology is used only for nefarious activities. Establishing a binary between the darknet and the regular internet – or the 'lightnet', according to the metaphor – suggests that technology such as TOR is inherently bad. For example, Moore and Rid (2016) try to separate the 'good' aspects of the darknet from the 'bad'. They distinguish between activities such as the quest for greater anonymity on the part of human rights activists and malicious (particularly criminal) activities. But this distinction cannot always be easily drawn, even when considering illegal activities and their impact. Of course, offering narcotics over the internet is illegal, yet it allows for more transparency through customer ratings and hence more secure buying of drugs. Thus, the impact of these pages might not be wholly negative; indeed, they might be highly beneficial for drug users. The notion of a darknet is thus deeply problematic, as is a uniformly negative assessment of this technology.

Finally, encryption is also negatively connoted in the context of ransomware. This term refers to programs that are used to attack a computer and encrypt all of its data. The victim must then pay a ransom in order to obtain the key to decrypt the data. Thus far, the most famous example has been 'WannaCry' in 2017. Most of the attacks occurred in India, but the German and Russian railway systems, the British National Health Service and universities around the world were also targeted (Mohurle and Patil, 2017). Here, encryption technology was deliberately used to launch a malicious attack. However, the debate over ransomware seems to be detached from the core theme of this book: namely, encryption, security and surveillance.

To conclude, the political character of encryption came to the fore most explicitly during the cryptowars. The state and activists – as well as commercial interests – have all been embroiled in debates about encryption and how these controversies transcend national boundaries, yet the roles of these actors are often ambiguous. Some commercial actors are in favour of stronger encryption while others want to see it weakened, and government agencies such as the FBI and the NSA often exhibit radically different opinions. This section also introduced the various loci where encryption has been debated and appears as a security technology. The focus of this book is on encryption as a security technology in the context of debates over surveillance and privacy. However, encryption is also crucial to blockchain technologies that have the potential to alter established social and economic orders and modes of organisation.

Conclusion: the politics of encryption

Nowadays, few people would deny the usefulness of encryption, for instance in securing medical devices. Hence, it is difficult to present the issue in simple for-or-against terms. Of course, strong positions are evident on both sides, but the real debate is over the degree to which encryption is implemented and the specific role it should play in the economy or with regard to privacy. For example, agencies such as the federal government, the NSA and the FBI have at various times held quite different opinions about encryption as a security technology. The controversies also revolve around the question of what is known. How certain are we that a particular encryption program provides security for our emails? How will we ever know how often encryption has actually obstructed law enforcement? These questions are at the core of encryption controversies. In this respect, this chapter has offered a further illustration of Barry's observation that controversies are never about a single issue; rather, they always imply a contest over what can be known and who or what is involved (Barry, 2012).

The politics of encryption comes to the fore on quite different planes. 'Traditional' politics are evident when we look at how governments have tried to implement escrow systems. Questions about technological standards and export regulations have long been at the heart of the politics of encryption. The activities of hackers and cypherpunks can be quite radical in the sense that they try to challenge the established power structure. But in a way the underlying principle follows a traditional understanding of politics in which questions concerning power and inequality are debated with respect to the state. On the other hand, the framework of the nation-state is also transcended in at least two ways: first, standards set in the United States are always de facto binding for the rest of the world, which means that other states cannot exercise sovereignty over the question of encryption; second, Zimmermann 'exported' encryption simply by putting it online. In the context of the global but US-centred internet, regulations governing the export of code simply cannot work. This topic will be discussed in greater detail in Chapter 7, with further reflection on what these observations mean for our political vocabulary.

In sum, it is apparent that security politics occurs in a variety of spaces and through a variety of practices. Though technology is often a crucial aspect of this, it must be understood as tightly interlinked with broader social questions.

Notes

1 At the time of writing, the NIST was in the process of selecting a 'post-quantum' encryption system. This should be available by 2022 (see NIST, 2019).
2 For the global economy and global communication, one standard is necessary. A 'private' encryption algorithm would be the most insecure type since no other party could have checked it for flaws. This is why so few encryption algorithms are available. Establishing a new one is a long-winded process. It would be difficult to contest the de facto US hegemony since the security of encryption relies on the fact that it needs to be tested by many people. I thank Malcolm Campbell-Verduyn for helping me to clarify this point.
3 MIT had been involved in the development of PGP, but as it had also been involved in the development of the patented algorithm (and because prosecuting a prestigious university would have be more difficult than going after an individual such as Zimmermann), a separate lawsuit against the institute was soon dropped, too (Diffie and Landau, 1998: 206; Schwenk, 2010: 32).
4 For an overview of the controversy from different perspectives, see: Barrett (2016), Burgess (2016) and Hennessey and Wittes (2016).
5 Gehl (2016) provides one of the few pieces of ethnographic research on people using the 'darknet' as an anonymous social network and thereby focuses not on criminal

activities but on the practice of searching for anonymity more generally. He shows that the search for anonymity need not be related to criminal activities, but constitutes a value in itself for many users.

Bibliography

Acharya, B. (2017). Breaking ranks with Asia: The case for encrypting India. *ORF*, 21 February. Available at: www.orfonline.org/expert-speaks/breaking-ranks-with-asia-the-case-for-encrypting-india/ (accessed 23 April 2019).

Apple Inc. (2016). Customer letter. Available at: www.apple.com/customer-letter/ (accessed 24 February 2016).

Aradau, C. (2017). Assembling (non)knowledge: Security, law, and surveillance in a digital world. *International Political Sociology* 4(1): 327–342.

Assange, J.*et al.* (2012). *Cypherpunks*. New York: OR Books.

Atzori, M. (2015). Blockchain technology and decentralized governance: Is the state still necessary? *SSRN Electronic Journal*. Available at: https://papers.ssrn.com/sol3/papers.cfm?abstract_id=2709713 (accessed 23 April 2019).

Barret, D. (2018). FBI repeatedly overstated encryption threat figures to Congress, public. *Washington Post*, 22 May. Available at: www.washingtonpost.com/world/national-security/fbi-repeatedly-overstated-encryption-threat-figures-to-congress-public/2018/05/22/5b68ae90-5dce-11e8-a4a4-c070ef53f315_story.html (accessed 2 April 2019).

Barrett, B. (2016). Don't be misled: The Apple–FBI fight isn't about privacy vs. security. *Wired*, 24 February. Available at: www.wired.com/2016/02/apple-fbi-privacy-security/ (accessed 23 April 2019).

Barry, A. (2012). Political situations: Knowledge controversies in transnational governance. *Critical Policy Studies* 6(3): 324–336.

Burgess, M. (2016). Apple and FBI: Department of Justice unlocks San Bernardino iPhone. *Wired*, 7 April. Available at: www.wired.co.uk/news/archive/2016-03/29/apple-fbi-unlock-iphone-5c-court-order-dropped (accessed 23 April 2019).

Burghardt, T. (2015). The US secret state and the internet: 'Dirty secrets' and 'Crypto Wars' from 'Clipper Chip' and ECHELON to PRISM. *Global Research*, 19 May. Available at: www.globalresearch.ca/the-u-s-secret-state-and-the-internet-dirty-secrets-and-crypto-wars-from-clipper-chip-to-prism/5357623 (accessed 23 April 2019).

Campbell-Verduyn, M. (ed.) (2017). *Bitcoin and beyond: Cryptocurrencies, Blockchains, and Global Governance*. Abingdon and New York: Routledge.

Carr, M. (2016). Public–private partnerships in national cyber-security strategies. *International Affairs* 92(1): 43–62.

Cerulus, L. (2017). EU encryption plans hope to stave off 'backdoors'. Available at: www.politico.eu/article/eu-encryption-plans-boost-power-of-europol-police-to-fight-cybercrime-backdoors-hack-phones-computers/ (accessed 1 May 2018).

Chaos Computer Club (2011). CCC | Chaos Computer Club analysiert Staatstrojaner. Available at: www.ccc.de/de/updates/2011/staatstrojaner (accessed 19 September 2018).

Coleman, G. (2011). Hacker politics and publics. *Public Culture* 23(3): 511–516.

Crocker, A. (2018). FBI admits it inflated number of supposedly unhackable devices. Available at: www.eff.org/deeplinks/2018/05/fbi-admits-it-inflated-number-supposedly-unhackable-devices (accessed 16 August 2018).

Curtin, M. (2005). *Brute Force: Cracking the Data Encryption Standard*. New York: Copernicus Books.

De Filippi, P. and Wright, A. (2018). *Blockchain and the Law: The Rule of Code.* Cambridge, MA: Harvard University Press.

Denning, D.E. (2001a). Afterword to 'The future of cryptography'. In: Ludlow, P. (ed.) *Crypto Anarchy, Cyberstates, and Pirate Utopias.* Cambridge, MA: MIT Press, pp. 103–104.

Denning, D.E. (2001b). The future of cryptography. In: Ludlow, P. (ed.) *Crypto Anarchy, Cyberstates, and Pirate Utopias.* Cambridge, MA: MIT Press, pp. 85–101.

Diffie, W. and Landau, S.E. (1998) *Privacy on the Line: The Politics of Wiretapping and Encryption.* 1st edition. Cambridge, MA: MIT Press.

Diffie, W. and Landau, S.E. (2007) *Privacy on the Line: The Politics of Wiretapping and Encryption.* Updated edition. Cambridge, MA: MIT Press.

Dooley, J.F. (2013). *A Brief History of Cryptology and Cryptographic Algorithms.* New York: Springer.

Dunn Cavelty, M. (2007). *Cyber-security and Threat Politics: US Efforts to Secure the Information Age.* CSS Studies in Security and International Relations. Abingdon and New York: Routledge.

DuPont, Q. (2014). Unlocking the digital crypt: Exploring a framework for cryptographic reading and writing. *Scholarly and Research Communication* 5(2): 1–8.

DuPont, Q. (2017). Experiments in algorithmic governance: A history and ethnography of 'the DAO', a failed decentralized autonomous organization. In: Campbell-Verduyn, M. (ed.) *Bitcoin and beyond: Cryptocurrencies, Blockchains, and Global Governance.* Abingdon and New York: Routledge, pp. 157–177.

EFF (2016). EFF DES cracker machine brings honesty to crypto debate. Available at: www.eff.org/press/releases/eff-des-cracker-machine-brings-honesty-to-crypto-debate (accessed 14 November 2018).

FinFisher (2018). Home page. Available at: https://finfisher.com/FinFisher/index.html (accessed 14 November 2018).

FISA (2015). *Annual Report to Congress.* Available at: http://fas.org/irp/agency/doj/fisa/#rept (accessed 28 April 2016).

Galloway, A.R. (2006). *Protocol: How Control Exists after Decentralization.* Leonardo. Cambridge, MA: MIT Press.

GammaGroup (2018). Home page. Available at: www.gammagroup.com/ (accessed 14 November 2018).

Ganz, K. (2018). *Die Netzbewegung: Subjektpositionen im politischen Diskurs der digitalen Gesellschaft.* Opladen, Berlin and Toronto: Verlag Barbara Budrich.

Gehl, R.W. (2016). Power/freedom on the Dark Web: A digital ethnography of the Dark Web social network. *New Media and Society* 18(7): 1219–1235.

Goitein, E. and Patel, F. (2015). What went wrong with the Fisa court? Brennan Center for Justice at New York University School of Law. Available at: www.brennancenter.org/publication/what-went-wrong-fisa-court (accessed 23 April 2019).

Hellegren, Z.I. (2017). A history of crypto-discourse: Encryption as a site of struggles to define internet freedom. *Internet Histories* 1(4): 285–311.

Hennessey, S. and Wittes, B. (2016). Apple is selling you a phone, not civil liberties. *Lawfare*, 18 February. Available at: www.lawfareblog.com/apple-selling-you-phone-not-civil-liberties (accessed 23 April 2019).

Herrera, G.L. (2002). The politics of bandwidth: International political implications of a global digital information network. *Review of International Studies* 28(1): 93–122.

Kahn, D. (1997). *The Codebreakers: The Comprehensive History of Secret Communication from Ancient Times to the Internet.* New York: Scribner's and Sons.

Kościelny, C., Kurkowski, M. and Srebrny, M. (2013). Public key infrastructure. In: Kościelny, C., Kurkowski, M. and Srebrny, M., *Modern Cryptography Primer.* Berlin and Heidelberg: Springer Berlin Heidelberg, pp. 175–191.

Kubitschko, S. (2015). The role of hackers in countering surveillance and promoting democracy. *Media and Communication* 3(2): 77.

Kubitschko, S. (2018). Chaos Computer Club: The communicative construction of media technologies and infrastructures as a political category. In: Hepp, A., Breiter, A. and Hasebrink, U. (eds) *Communicative Figurations: Transforming Communications in Times of Deep Mediatization.* Cham: Springer International Publishing, pp. 81–100.

Lavabit (2016). Home page. Available at: http://lavabit.com/ (accessed 21 April 2016).

Levy, S. (1994). Cypher wars: Pretty Good Privacy gets pretty legal. Available at: http://encryption_policies.tripod.com/industry/levy_021194_pgp.htm (accessed 9 February 2015).

Lewis, J.A. *et al.* (2017). *The Effect of Encryption on Lawful Access to Communications and Data.* Available at: https://csis-prod.s3.amazonaws.com/s3fs-public/publication/170203_Lewis_EffectOfEncryption_Web.pdf?Gqb5hXxckXykb3WAphuoVHrHfDT wuFkN (accessed 4 September 2018).

Lyon, D. (2014). Surveillance, Snowden, and Big Data: Capacities, consequences, critique. *Big Data and Society* 1(2). Available at: https://journals.sagepub.com/doi/full/10.1177/2053951714541861 (accessed 2 April 2019).

Marquis-Boire, M. (2013). *For Their Eyes Only: The Commercialization of Digital Spying.* Available at: https://citizenlab.org/2013/04/for-their-eyes-only-2/ (accessed 24 June 2015).

Meinrath, S.D. and Vitka, S. (2014). Crypto War II. *Critical Studies in Media Communication* 31(2): 123–128.

Mohurle, S. and Patil, M. (2017). A brief study of Wannacry threat: Ransomware attack 2017. *International Journal of Advanced Research in Computer Science* 8(5): 1938–1940.

Moore, D. and Rid, T. (2016). Cryptopolitik and the darknet. *Survival* 58(1): 7–38.

NIST (2019). Post-quantum cryptography. Available at: https://csrc.nist.gov/projects/post-quantum-cryptography (accessed 15 April 2019).

Oppliger, R. (2005). *Contemporary Cryptography.* Norwood, MA: Artech House.

Page, S. (2016). Ex-NSA chief backs Apple on iPhone 'back doors'. *USA Today*, 7 April. Available at: www.usatoday.com/story/news/2016/02/21/ex-nsa-chief-backs-apple-iphone-back-doors/80660024/ (accessed 23 April 2019).

Parsons, C. (2016). Christopher Parsons on the RCMP's BlackBerry encryption key. Available at: https://citizenlab.ca/2016/04/christopher-parsons-blackberry-encryption-key/ (accessed 14 November 2018).

Poulsen, K. (2013). Edward Snowden's email provider shuts down amid secret court battle. Available at: www.wired.com/2013/08/lavabit-snowden/ (accessed 13 November 2018).

Saco, D. (2002). *Cybering Democracy: Public Space and the Internet.* Electronic Mediations. Minneapolis: University of Minnesota Press.

Salomon, D. (2006). *Elements of Cryptography.* London: Springer-Verlag.

Schneier, B. (1996). *Applied Cryptography: Protocols, Algorithms, and Source Code in C.* New York: Wiley.

Schoen, S. (2010). BlackBerry bans suggest a scary precedent: Crypto Wars again? *Electronic Frontier Foundation*, 8 April. Available at: www.eff.org/de/deeplinks/2010/08/blackberry-bans-suggest-scary-precedent (accessed 23 April 2019).

Schulze, M. (2017). Clipper Meets Apple vs. FBI: A comparison of the cryptography discourses from 1993 and 2016. *Media and Communication* 5(1): 54.

Schwenk, J. (2010). *Sicherheit und Kryptographie im Internet: Von sicherer E-mail bis zu IP-Verschlüsselung.* Wiesbaden: Vieweg + Teubner.

Singh, S. (2000). *The Code Book: The Science of Secrecy from Ancient Egypt to Quantum Cryptography.* New York: Anchor Books.

Singh Guliani, N. (2016). 4 problems with creating a 'commission on encryption'. Available at: www.aclu.org/blog/privacy-technology/internet-privacy/4-problems-creating-commission-encryption (accessed 1 May 2018).

Swire, P.P. and Ahmad, K. (2012). Encryption and globalization. *Columbia Science and Technology Law Review* 13: 416–481.

Taylor, P.A. (2005). From hackers to hacktivists: Speed bumps on the global super-highway? *New Media and Society* 7(5): 625–646.

TOR Project (2018). Overview. Available at: www.torproject.org/about/overview.html.en#overview (accessed 19 September 2018).

UNHCR (2015). *Report of the Special Rapporteur on the Promotion and Protection of the Right to Freedom of Opinion and Expression, David Kaye.* A/HRC/29/32. Geneva: UNHCR.

Usher, S. (2013). Saudi Arabia blocks Viber messaging service. *BBC News*, 6 June. Available at: www.bbc.com/news/world-middle-east-22806848 (accessed 23 April 2019).

Walters, W. (2014). Drone strikes, dingpolitik and beyond: Furthering the debate on materiality and security. *Security Dialogue* 45(2): 101–118.

Yadron, D. (2015). Obama sides with Cameron in encryption fight. *Wall Street Journal*, 16 January. Available at: http://blogs.wsj.com/digits/2015/01/16/obama-sides-with-cameron-in-encryption-fight/tab/print/?mg=blogs-wsj&url=http%253A%252F%252Fblogs.wsj.com%252Fdigits%252F2015%252F01%252F16%252Fobama-sides-with-cameron-in-encryption-fight%252Ftab%252Fprint (accessed 6 July 2016).

Zangrilli, A. (2012). 30 years of public-key cryptography. Available at: http://technology.findlaw.com/legal-software/30-years-of-public-key-cryptography.html (accessed 23 April 2019).

Zittrain, J.L.et al. (2016). *Don't Panic: Making Progress on the 'Going Dark' Debate.* Cambridge, MA: Berkman Center.

5 Encryption and national security

In a statement to the Subcommittee on Information Technology in 2016, Amy Hess, then executive assistant director of the FBI's Science and Technology Branch, discussed the challenges to law enforcement posed by new technologies, especially encryption. In her testimony, she warned of the dangers of strong encryption, of the state 'going dark'. She related a story about a trucker who 'imprisoned [his girlfriend] within his truck, drove her from state to state, and physically and sexually assaulted her along the way' (Hess, 2016; see also Comey, 2015). Then she pointed out that the trucker had been caught only because it had been possible to 'extract that data' on his phone. In the end, 'the trucker was convicted' and 'sentenced to life in prison' (Hess, 2016).

This is only one of many stories that are used to highlight the need for access to encrypted data.[1] Hess's testimony exemplifies how the state considers encryption as a threat and an obstacle to law enforcement, and how vividly this threat is depicted. However, later in this chapter I will show that an alternative security narrative depicts encryption as a *source* of higher security. Encryption thus figures as both a threat and a solution to security concerns. Hence, encryption is a contested issue: a potential obstacle to security, but also a potential tool for improving it. While many governments, especially that of the United States, consider strong encryption a threat to the conduct of law enforcement, civil rights associations present it as an essential tool for securing free speech and privacy. The role of business is ambiguous, too. Some ICT companies are perceived as threats, since they cooperate with the state and therefore make mass surveillance possible. Meanwhile, others offer strong encryption to their customers, thereby allowing a higher level of privacy. Such initiatives are viewed critically by secret services.

The focus of this chapter is to understand how 'security' plays out in the debates on encryption and on the underlying ideas of security that shape these discussions. Investigating the concept of security helps us understand the assumptions on which discussions of encryption are built. This concept of security does not refer to an analytical concept invoked by scholars or to some ontological notion of what it 'really' looks like. I will not propose my own concept of security and then analyse security practices in its light. Rather, I will examine the various ideas that prevail among experts, politicians and among mass media. The aim is to reconstruct how dominant assumptions about the role of encryption for security play out.

Methodologically, I performed a fine-grained analysis of primary sources (listed in the Appendix) by reconstructing the prevailing assumptions concerning security, threats and solutions (see Chapter 2). Explicating the assumptions underlying the encryption discourse not only enables a more concise description of the discourse but also helps to clarify how these assumptions contribute to certain effects. The results are presented in the form of two narratives that will highlight the prevailing ideas, assumptions and arguments that shape the debates about encryption technology and security. The citizen controlling her own data, on the one hand, and the state conducting mass surveillance, on the other, appear as central threats in the respective accounts. I will show how security is individualised by presenting encryption as a tool to be implemented by single users in order to protect themselves against the state. Through this reconstruction of security narratives I will demonstrate that the constitutive distinction between public and private is not clear cut in the case of encryption. The role of 'private' actors, such as global ICT companies, is rather ambiguous, and not all companies can be subsumed with reference to the same set of interests. This chapter highlights the individualisation of security, which becomes something that is enacted on the private level. For activists, security is not a good that is provided by the state; instead, it is something to which every user must attend on their own. Controversies thus challenge established political practices in the sense of Dewey's concept of publics (see Chapter 3). Technological controversies thus not only concern the implementation and spread of technology, but also challenge the role of the state as the prime provider of security.

Furthermore, the investigation of encryption as a security technology highlights the ambiguous character of encryption and how technological contestation is embedded in threat narratives. Analysis of these threat narratives reveals that the distinction between public and private – and, by extension, that between privacy and security – does not hold much analytical value. This is partly due to the fact that the debates about encryption cannot be organised around a simple for/against dichotomy. Accordingly, I argue throughout this book that controversies about encryption cannot be understood by presenting them in the form of 'the public versus the state' or 'public versus private'. Instead, various forms of publicness emerge or – as this chapter shows – their emergence is complicated by the prevailing security narratives. Before presenting these narratives, however, the next section offers a reconstruction of 'risk' as the underlying logic of the encryption discourse, which anticipates the important themes of the individualisation and diffusion of security in the encryption debates.

Risks in cyberspace

Research on cybersecurity often characterises the phenomenon of cyberattacks as one that transcends the established categories in which traditional Security Studies operates. Cyberattacks are said to exceed levels of analysis such as individual, national and global. The debate mostly revolves around the novelty of these attacks and the authenticity of scenarios that attempt to capture them (Arquilla and Ronfeldt, 2001; Rid, 2013). However, here I am more interested in the assumptions and narratives about security that prevail on the level of practice (see Guzzini, 2000: 161). The aim is to reconstruct notions that can be found among experts, politicians and in the media. Hence, there is no attempt to determine if this is a 'correct' assessment of encryption technology. Rather, the goal is to establish how assumptions about digital technology and security shape the controversies revolving around encryption.

First, notions of security that are dominant on the level of practice need to be considered against certain ideas about digital technology. One prevalent feature of the discourse on internet security – in academia as well as in the broader public sphere – is the idea of the 'digital era', which is said to be quite distinct from previous times. This general assumption also features in the discourse on encryption in both Germany and the United States. For instance, President Obama addressed the issue of encryption when he tried to recruit IT experts to work for the government, linking the notion of a special 'moment in history' with the specific demands this was posing for the economy and national security:

> Look, we are at a moment in history where technology, globalization, our economy is changing so fast. And this gathering, South by Southwest [an annual arts and music festival in Austin, Texas], brings together people who are at the cutting edge of those changes. Those changes offer us enormous opportunities but also are very disruptive and unsettling. They empower individuals to do things that they could have never dreamed of before, but they also empower folks who are very dangerous to spread dangerous messages.
>
> (Obama, 2016)

For Obama, changing technology and changing social factors were having an impact on individuals, so no one could avoid being affected by these new technologies, for better or for worse (see also Comey, 2015; *Wired*, 2015). On its web page on cybercrime, the BKA (Bundeskriminalamt – the German Federal Police) similarly states that 'modern societies are networked [*vernetzt*]', but alongside the 'positive possibilities of the use of the internet, negative side-effects occur as well: cybercriminals are offered multiple opportunities to attack' (BKA, 2016).[2] The notion that we live in a special time in which technology shapes the development of society can also be found in official documents, such as a United Nations report on the protection of freedom of speech that deals primarily with the challenges presented by networked technology (UNHCR, 2015). In these texts technology is depicted not merely as a means to an end; rather, it is presented as a fundamental characteristic of a certain kind of society. Remarks about 'a digital era' or the 'Snowden era' are made to highlight that technology is ubiquitous and ever-present, with Snowden repeatedly invoked as having inaugurated a new era (e. g. Mandau, 2014a; Steinschaden, 2014a; for a reference in the academic literature, see Bauman *et al.*, 2014). Tim Stevens speaks in this context of a 'problematic periodization in cyber security' in which everything that relates to cybersecurity is framed as radically new and part of a 'revolution', rather than merely an aspect of technological evolution (Stevens, 2016: 70). Similar discursive features can be observed in the encryption debates.

This observation testifies to the concept of the technological society (Barry, 2001) as a society that is defined through its preoccupation with science and technology for describing and solving societal issues. According to Barry, the decisive criterion for describing contemporary society is not the presence or absence of digital technology but the role it plays in thinking about society. The conception of a technological society thus relies on a certain self-description of society of which the notion of a digital era is one part.

This is highly relevant to this study because the notion of a digital era is invariably linked to the advent of specific threats. Debating security and encryption always occurs against the backdrop of a certain age in which digital technologies determine the possibilities of (in)security. This is best observed in the mass media discourse:

> In many companies, notebooks and smartphones have long replaced filing cabinets. On the small devices many secrets can be saved – but also can be stolen in no time at all.
>
> (Martin-Jung, 2012)

Importantly, this extract from a 2012 article in a German newspaper describes changes that can be observed in all kinds of companies, not just a few cutting-edge corporations. As in Obama's declaration, there is some ambivalence in the assessment of this technology. Digital devices might be useful, but they allow for easier theft from both companies and individuals. While security is obviously a concern for nation-states it should not be seen solely from a military perspective, because it is a problem that affects individuals, too. Technologies spread and devices become smaller, which is understood as a precondition for increased danger. The very fact that new technologies allow ever more data to be saved increases the potential damage if just one such device is stolen. The idea that we live in a digital era of ubiquitous technology is linked to the emergence of new threats. Possibilities for attacks that emerge in relation to the new technologies that define the digital era mushroom. Living in the digital era thus entails living with a distinct set of insecurities.

The vulnerability of critical infrastructure to attack is one of the predominant threat scenarios in which encryption is seen as a tool that may provide better security. This has been a common theme since the 1990s in the United States, as Dunn Cavelty (2007: 5) has shown. Since then, attacks on infrastructure have often been linked to cyberthreats (Dunn Cavelty, 2007: 92). For example, the German popular novel *Blackout* depicts a cyberattack on the

European electricity network and the catastrophic consequences this has for society (Elsberg, 2012). The novel was widely discussed in the mass media and has subsequently been translated into 15 languages (Wikipedia, 2018). It was generally received as a realistic depiction of the vulnerability of critical infrastructure and the potential effects of a cyberattack (Kutter and Rauner, 2012; Lange, 2012), and shows how conceptions of low-probability but high-impact situations in the context of internet security have become dominant in public discourse. Although many experts insist that such scenarios are unrealistic, demands to prevent them are still frequently made. Consider the following statements by Eugene Kaspersky in a German newspaper:

> Improbable? To be sure. But it is sufficient for this scenario to happen once. Even then malware can spread around the globe exponentially and hit everybody, even the country from which the attack started. Everybody, that means: private users of computers, companies, government and whole states.
>
> (Kaspersky, 2012)

Images of networked, ubiquitous technology go hand in hand with images of threats that can occur anywhere, at any time and can target anything. Kaspersky's article reveals how ideas about security that are familiar from academic discourse can also be found at the level of practice. It reflects the logic of the precautionary principle, according to which preparations must be made for unlikely events if they have the potential to cause a great deal of damage (Aradau and Van Munster, 2007; Beck, 1986).

The examples quoted above show how different threats, emerging from a variety of sources, targeting various objects and actors, are depicted in the debates revolving around encryption. The targets may be individuals (criminals), companies (espionage) or nation-states (destruction of infrastructure). Security is imagined as characterised by these attacks that may occur at any time, target anyone and cause extensive damage, notwithstanding their infrequency or unlikelihood. Thus, precautionary measures must be taken.

This rationale of precaution leads to a focus on technologies that may result in greater security. Encryption is a prime tool for providing protection against espionage and securing sensitive data or infrastructure against attacks from other countries or organised crime. The link to the digital era theme is apparent in one of security expert Bruce Schneier's blog posts: 'In today's world of ubiquitous computers and networks, it's hard to overstate the value of encryption. Quite simply, encryption keeps you safe' (Schneier, 2016). However, neither encryption nor any other technology is presented as providing *full* protection. Rather, encryption's effect is perceived as much more ambiguous. For instance, it only works if it is properly implemented, updated and not compromised. This view appears especially in texts written by experts, who insist on the complexity of encryption. It is rarely presented as the ultimate tool for complete security. Instead, the suggestion is that it makes attacks much more difficult, time-consuming and therefore costly. Ever-present threats that shape the digital era cannot be completely neutralised. Good technology merely decreases the probability of an attack or makes the attack more costly. This theme of a reduction of risk via encryption prevails not only among experts but also in the mass media. One example is drawn from an article in the *Washington Post* that deals with the relationship between the FBI and Apple:

> Encryption, while not impervious to targeted surveillance, makes it much more difficult to read communications in bulk as they travel the Internet. The NSA devotes substantial resources to decoding encrypted traffic, but the work is more targeted and time consuming, sometimes involving hacking into individual computers of people using encryption technology.
>
> (Timberg *et al.*, 2013)

Sandro Gaycken, an internet security expert from Germany, makes the following comment about encryption and the threat from secret services: 'A sufficiently strong encryption of all communication is, for example, a good and efficient way to make foreign secret services lose interest in mass surveillance. Decryption on a mass scale is outside the operational rationality [*Betriebsrationalität*] of secret services' (Gaycken, 2014). Mass surveillance is viable only when users do not use any tools to protect themselves. If relatively simple tools were used to increase the cost of an attack, it would be impossible. Markus Mandau discusses encryption in the specialist magazine *CHIP* in a similar way: 'This [kind of encryption] does not protect against targeted attacks, but makes it impossible to collect all data traffic via submarine cables' (Mandau, 2014b).

Therefore, experts agree that encryption is a good tool for increasing security, but full protection cannot be guaranteed. Encryption systems differ in strength, and even though the underlying systems might be powerful, they still need to be implemented correctly. The resulting complex system allows for a variety of weak spots that may be targeted. While good systems provide better security, any encryption system can be breached with enough time and money. The whole logic of how to increase security is thus based on a cost-efficiency logic. The cost of potential damage needs to be balanced against the cost of implementing counter-measures. In this sense security is conceptualised in *economic* terms. Indeed, the concept of low probability/ high damage is familiar from the risk literature in Security Studies (for an overview, see: Petersen, 2012). According to this logic, one must consider not only the likelihood of a threat but also its potential (large-scale) damage (Aradau and Van Munster, 2007). Such a security discourse is characterised by references to unpredictability and the ubiquity of threats (Amoore, 2011; Amoore and De Goede, 2005; Daase and Kessler, 2007). A threat cannot be erased, but the impact of an attack can be reduced. In the debates on encryption, insecurity is depicted in terms of 'risks' that must be reduced, rather than 'threats' that must be eliminated. The focus on (incalculable) risk also leads to an individualisation of security: 'Conjoined with the individualisation of risk is the logic of individual responsibility for containing them' (Krahmann, 2011: 362). A dominant logic of risk thus also speaks to the individualisation of security – a topic that will be explored further when discussing the narrative of the state out of control.

In the context of cybersecurity, there is a sense that we are 'trapped in an accelerating present, cut off from history and with few options for controlling the future' (Stevens, 2016: 93). This analysis links the theme of living in a distinct era with the kinds of threat that emerge due to new technologies. The latter are developing at an ever faster pace, and security measures cannot keep up with them (Stevens, 2016: 89, 115). Thus, risk and uncertainty about the future shape any discussion about encryption and security. Based on my reconstruction, we can note that a logic of risk prevails in the security discourse on encryption. Leaving aside that discourse's accuracy, we can see how security politics are 'sung into existence' (Smith, 2004). Narratives about the possibility of cyberattacks shape how security technology is perceived. The resulting picture is one of a society that is moulded by its use of increasingly widespread, sophisticated technology, and consequently must address diffuse insecurities.

The double role of encryption

Narrating encryption and security relies on assumptions about risk and uncertainty as decisive features of networked security. However, we can distinguish between two narratives in which a certain threat, its solution and the main actors are embedded. Encryption plays out very differently in these two narratives, which is why we should examine them more closely. It is imagined as both a problem and a solution. To make matters even more complicated, different levels – the economy, the state and citizens – form part of these narratives and structure how threat scenarios play out within them. I hope that the reconstruction of these narratives will provide a better understanding of how encryption is a contested issue. The two opposing narratives bring to the fore the struggles about the role of encryption for security.

The 'state in control' narrative is the view of encryption from a nation-state's perspective. Two concerns are paramount here: national security and the national economy (see Table 5.1). On the one hand, the security of citizens is threatened by criminals, organised crime and terrorists. Encryption plays a crucial role in this threat scenario as it allows suspects to communicate secretly and without fear of interception by law-enforcement agencies. It hampers efforts to monitor and identify suspects and separate the good from the bad. Thus, surveillance measures that are seen as a means to increase security are complicated or even made impractical. On the other hand, the economy is also in need of protection, and a good economy in the digital age, so the narrative goes, is only possible with strong encryption. Thus, encryption cannot be completely forbidden as it is also necessary for economic security. Hence, claims to regulate encryption need to be balanced against claims for the need of strong encryption for economic reasons. To make matters even more complicated, 'the economy' is far from a homogeneous category in the discourse. While many companies rely on strong encryption for e-commerce, others, including Google, need unencrypted data in order for their business models to function. This is because only unencrypted data provides the information that is needed for targeted advertising. Furthermore, companies offering end-to-end encryption by default are viewed as threats to the state, because strong encryption allegedly hampers law enforcement. Nevertheless, in the aftermath of the Snowden revelations, many companies started to offer encryption by default to satisfy their customers' demands for more privacy and to bolster trust in their products. Thus, ICT companies that offer strong encryptionare now widely considered as a threat from a state perspective.

Table 5.2 summarises the second narrative, in which the state features as the main threat from the citizens' perspective. In this case the object of protection is the private sphere and, by extension, democracy itself (see below). The threat is identified as coming mainly from the state and its surveillance activities. ICT companies collecting data are also considered threats, either because their collection of data for commercial purposes is considered to be an invasion of privacy or because they cooperate with the state in its surveillance efforts. Encryption is thus presented as a means of increasing security, which is to say protecting privacy. Hence, companies may be seen not only as threats but also as potential enablers of privacy, provided they offer strong encryption. Companies' activities are interpreted as a problem in the first narrative, but considered a necessity to protect privacy in the second. The effect of this threat scenario is that activists demand encryption by default from companies. Furthermore, open source encryption software and other applications to protect privacy are favoured by activists to protect privacy. Thus, publicness emerges in the debate over the role of companies and the extent to which the state can still be trusted. Encryption therefore features as only one tool in a broader initiative to strengthen privacy.

Table 5.1 The state in control

Object of protection	Threat	Means of protection	Effect on encryption
National security	Citizens as possible criminals, terrorists, paedophiles, drug dealers	Surveillance	Restrict and control use of encryption
National economy	Cyberespionage, cyberattacks (harming commerce)	Provide strong digital infrastructure	Permit encryption

Table 5.2 The state out of control

Object of protection	Threat	Means of protection	Effect on encryption
Private sphere and, by extension, democracy	Surveillance by state and companies	Strong encryption	Foster open source encryption, demand that companies offer stronger encryption by default

As may be deduced from Tables 5.1 and 5.2, the role of the economy is quite complex and plays out in myriad ways. Hence, I will first describe the different ways in which references to economic arguments are made and the various roles companies play in this discourse. In other words, I will describe their ambiguous character before turning to the first narrative.

The economy as both a threat and an object of protection

One can hardly speak about 'the economy' as a unitary actor in the case of encryption. Myriad interests play out in the politics of encryption. The economy is both a threat and an object of protection. Protecting it by increasing internet security is considered to be imperative. The economic threat of insecure technology and hacking attacks drives many state actors' internet security measures. But certain companies are also perceived as threats in the context of encryption technology: the more consumers apply tools that offer strong encryption, the more the state fears losing access to communication by criminals. To make matters even more complicated, companies are also assisting the state in monitoring the general public. Economic considerations have always been present when debating encryption. A strong economy in the twenty-first century needs good internet security to flourish, and encryption is considered to be central to achieving this. Encryption thus becomes a cornerstone of a strong economy. Consider this statement from the then director of the FBI, James Comey:

> [E]lectronic communication has transformed our society, most visibly by enabling ubiquitous digital communications and facilitating broad e-commerce. As such, it is important for our global economy and our national security to have strong encryption standards. The development and robust adoption of *strong encryption is a key tool to secure commerce and trade*, safeguard private information, promote free expression and association, and strengthen cyber security.
>
> (Comey, 2015; emphasis added)

The economy is the focal point of this discussion. Strong encryption is needed because the economy relies on it. This sentiment prevails in the United States but also in the German discourse, and all actors refer to it. One of the starkest statements on the subject can be found in a speech that Jeh Johnson, from the Department of Homeland Security, delivered to an audience of cybersecurity experts in 2015:

> Two weeks ago I was in Beijing [...] Though we have sharp differences with the Chinese Government, particularly when it comes to the theft of confidential business information and proprietary technology through cyber-intrusions, we and the Chinese recognize the need to make progress on a range of cyber-related issues. *As the two largest economies in the world, the US and China have a vested interest in working together to address shared cyber-threats*, and making progress on our differences.
>
> (Johnson, 2015; emphasis added)

The need for better internet security thus makes even cooperation with adversaries desirable. One specific aspect of this debate has come to the fore ever since the 1990s: imposing domestic legal constraints on encryption would simply lead actors to use foreign products. Criminals, especially, would not hesitate to use illegal foreign software in order to communicate securely. Thus, banning encryption in one country would have no impact on security. In the United States this argument is made with the added twist that forbidding strong encryption would not only be useless but would actively harm the US economy. For example, an article published in the *Washington Post* in 2016 declared:

> Moreover, proponents of encryption point out that numerous countries and groups have developed their own products and services, meaning anti-encryption policies will only hurt the competitiveness of US companies without providing access to a great deal of suspect communications.
>
> (Segal and Grigsby, 2016)

This argument is naturally not made in Germany as it is the country that profits most from strong regulation in the United States. After the United States, Germany is the second-largest producer of encryption software – a fact that is not explicitly translated into an argument against the regulation of encryption. However, the protection of the economy is a dominant theme in German discourse, too. For example, on the BKA website encryption is discussed in relation to its crucial importance for the economy (BKA, 2016). Similarly, the Bundesamt für Sicherheit in der Informationstechnik (BSI), the central German internet security institution, tends to emphasise protection of the economy (BSI, 2017). A more recent issue is the appearance of ransomware (see Chapter 4). Here, encryption technology is clearly perceived as a threat, but in a specific way – namely, as part of a cyberattack that might target companies, critical infrastructure or individuals. The WannaCry attack in 2017 was widely reported at the time (e.g. Balser *et al.*, 2017; Weaver, 2017; Perlroth and Sanger, 2017), and again the economy was cast as the main object of protection.

In sum, the focus when discussing encryption is usually on how it can protect the economy and/or on how strong regulation would harm innovation. These themes appear in all kinds of texts, and similar arguments are advanced by all of the actors who debate encryption. Even the civil rights group EFF sometimes focuses on the economic value of encryption (e.g. Timm, 2013), which shows the hegemonic character of this perspective.

In contrast to the theme that the economy must be protected, the role of companies is discussed in relation to how the data they collect is used by state agencies. In this respect, companies such as Google are characterised as potential enablers of mass state surveillance. In the wake of Edward Snowden's revelations, it was widely reported that leading IT and social media companies had cooperated with the secret services (e.g. Hintz and Dencik, 2016: 9; Berners-Lee, 2013; Lyon, 2014). As a result, Google, Facebook and Apple all allegedly lost customers. Companies suffer severe consequences whenever customers lose faith in them. The effect of this negative depiction led these companies to consider new technologies that would prevent future cooperation with state surveillance. The view that firms such as Microsoft, Apple and Google need to regain the trust of their customers in order not to endanger their business models remains as prevalent as ever. And this is where encryption comes into play, as Jacob Steinschaden, from the German organisation netzpiloten.de, commented in 2014:

> Almost a year and a half has passed since whistleblower Edward Snowden started to disclose the surveillance affair revolving around the NSA and the British GCHQ along with cooperating journalists. A few months later a study by the Cloud Security Alliance, the members of which include Microsoft, Google, Cisco, PayPal and Amazon, among others, shows that the scandal will cost the internet industry in the US\$35–45 billion over the next three years, since private and business customers might lose trust in cloud services. Since then, IT giants have been seeking to gain back this trust – and the means for that is called encryption.
>
> (Steinschaden, 2014a)

Encryption is seen as the core security and privacy tool that will enable companies to regain their customers' trust. Customers who fear tech companies' cooperation with the state might be persuaded to continue using their services if strong encryption is offered. If companies offer end-to-end encryption by default, the data cannot be read when it is transmitted. If the user is the only person who has the key for decryption, the company cannot pass on the data, even if the state tries to force it to do so. Furthermore, criminals will have a difficult time hacking into communications. Thus, tech companies have started to offer stronger encryption to their customers, as has been widely reported in the mass media. The broader discursive dynamics are illustrated in the following passage from the *Washington Post*:

Microsoft is moving toward a major new effort to encrypt its Internet traffic amid fears that the National Security Agency may have broken into its global communications links, said people familiar with the emerging plans. Suspicions at Microsoft, while building for several months, sharpened in October when it was reported that the NSA was intercepting traffic inside the private networks of Google and Yahoo [...] Though Microsoft officials said they had no independent verification of the NSA targeting the company in this way, general counsel Brad Smith said Tuesday that it would be 'very disturbing' and a possible constitutional breach if true.

(Timberg, Gellman and Soltani, 2013)

This article was published in November 2013, about six months after Snowden went public with his revelations. Microsoft is presented as uncertain whether the NSA had broken into its system. Secrecy about state and company actions increases uncertainty, which is a prevailing theme in this discourse. But the article also suggests that it is Microsoft's intention to provide better encryption, which 'stands to complicate surveillance efforts' (Timberg, Gellman and Soltani, 2013).[3] This theme of uncertainty about Microsoft's role ties in perfectly with the partially negative and often ambiguous depiction of Edward Snowden himself (Kuehn, 2018; Branum and Charteris-Black, 2015). In addition to widespread questioning of his motives, one strategy that was employed to dismiss his claims was to insist that there was no hard evidence of mass surveillance (Schulze, 2015). Uncertain knowledge thus plays a crucial role in the discourse and serves to legitimise companies' actions.

This strategy of legitimisation also pits companies against the state. In his analysis of the British and US discourse on encryption, Einar Thorsen asserts that 'computer companies implicated by the Snowden leaks were getting involved in public discussion about cybersecurity – to delegitimise mass surveillance and legitimise their own intelligence gathering about its users, usually expressed as highlighting their commitment to privacy' (Thorsen, 2017: 16). Offering encryption is seen as a tool for gaining legitimacy as a trustworthy company. In this context, implementing encryption is a means by which the company may distance itself from state actions that are perceived as illegitimate. This tactic makes the company's role more ambiguous: it may still be collaborating with the state, but it is also providing security against the state. The ambiguous role of companies complicates activists' resistance efforts as well as any kind of 'publicness' (see Chapter 3). Challenging global companies becomes more difficult if the target shifts and companies present themselves as allies against the state and its invasion of privacy.

A further remark on company–customer relations is needed to complete the picture. Problems with implementing encryption are also debated from the user's perspective. Strong encryption would make services such as Dropbox much more costly, as one German newspaper article observes (Mandau, 2014b). Moreover, encryption would make many applications 'less fun' (Berners-Lee,

2013). Especially in the expert statements that I analysed, concerns about usability were a prominent theme. Searching one's inbox, for example, would become slower and more complicated. Furthermore, as mentioned above, companies often rely on unencrypted data. Google, for instance, needs it to provide user-specific advertisements (and hence increase profits), so providing encryption must be balanced against economic interests. Microsoft, WhatsApp and Apple all provide quite strong encryption. Thus, major global companies do indeed provide strong privacy tools. However, in the newspaper articles that were analysed, the general sentiment is that such companies act primarily to increase their revenues. Their chief interest is profit rather than privacy.

The ambiguous role of companies described above fits the broader pattern of a political economy that is shaped by the commodification of data, which may be termed 'datafication' or 'data capitalism'. Sarah Myers West defines the latter term as follows:

> In using the term *data capitalism*, I aim to describe consequences of the turn from an e-commerce model premised on the sale of goods online to an advertising model premised on the sale of audiences – or, more accurately, on the sale of individual behavioral profiles tied to user data.
>
> (West, 2017: 4)

The user becomes the producer of data, which fosters not only a specific political economy but also modes of surveillance. Data capitalism relies on tracking users and collecting their data traces. Companies and state actors both rely on the same technology for surveillance. This 'internet–industry complex' encompasses state actors and private companies and produces a specific political economy (Flyverbom *et al.*, 2017). In addition, surveillance practices are often justified by reference to diffuse threats, such as terrorism, and accordingly mushroomed after 9/11 (Fuchs, 2013). Public and private actors often cooperate in these surveillance practices. Not only do secret services often rely on data collected by internet companies, but private companies also develop and sell surveillance technologies. The encryption discourse is embedded in this broader development. It is noticeable how discussions of regulation (of encryption) are lined to broader narratives on how to protect a political economy that relies on free access to and collection of data. Most importantly for the overall discussion on the issue of the public, explicating the ambiguous role of companies reveals that the dichotomy of public/private does not hold much analytical or normative value. Companies occupy multiple roles in the discourse, and the power of commerce is crucial for encryption debates. Yet, their role cannot be simply assigned to the side of activists or the state. As was highlighted in Chapter 2, though *public* and *private* are not clearly distinct realms, the roles of 'public' and 'private' actors form part of encryption controversies. This will become more apparent as the two main security narratives are presented below.

The state in control

We have already seen that the discourse on digital security is built on the idea that full protection is impossible and the only hope thus lies in increasing the cost of an attack. Security encompasses national and individual security as well as – and often most importantly – economic security. The crucial roles that companies, innovation and economic strength play in national security is a core part of this discourse. However, in the narrative of the 'state in control', companies are also an obstacle to security when they try to offer stronger encryption. Encryption that allows the user to have sole control over her data is deemed a significant threat, since this can be said to hinder law enforcement. As I show in this section, this ultimately means that too much privacy becomes a threat because encryption supposedly prevents the state from remaining in control. This section will unpack this narrative of why – from a state perspective – encryption is a problem for security.

'Going dark'

Throughout the twentieth century, encryption was presented as a concern for law enforcement on the basis that overly strong encryption would prevent the police and security services from gaining access to suspects' data. Today, companies that offer end-to-end encryption are perceived as the greatest threats. Those that provide stronger encryption supposedly hamper law enforcement because they enable complete privacy, and therefore prevent state agencies' access to data. The fear is that criminals, and especially terrorists, will use these secure means of communication for their own ends. This theme was evident as early as the 1990s (Denning, 2001). However, especially since 9/11, cyberthreats have been linked to terrorist attacks, whereas previously foreign states were seen as the main cause of concern (Dunn Cavelty, 2007: 103). In the aftermath of the terrorist attacks in Paris in 2015, arguments about encryption as a tool that empowers terrorists and hampers law enforcement intensified. From the state perspective, encryption is presented as an obstacle to providing security. The argument is that measures such as surveillance and interception of communications are made more difficult – or outright impossible – if strong encryption is used. The question is whether the state should have access to all encrypted communication, given that criminals might use strong encryption. As a 2015 article in the *Washington Post* notes:

> [The terrorist] attacks in Paris have sparked fresh debate about the risks and rewards of encrypted communications in smartphones and other devices and whether law enforcement should have what's called 'extraordinary access' in pursuit of criminals and terrorists. The tech companies say customers want to protect privacy. But encryption also can protect the communications of terrorists and other criminals.
>
> (Editorial Board, 2015)

Here, a link is made between companies interested in the privacy of their users and the problem their encryption initiatives pose for security. Although the article admits that it is not really clear if encrypted tools were used in the Paris attacks, the theme of encrypted communications and their hindrance of law enforcement was prominent in the months after these incidents, even leading British Prime Minister David Cameron to suggest banning encryption (Yadron, 2015). Although this suggestion met with fierce resistance, the direction of the debate was mostly about the perils of encryption. The theme that it poses a significant obstacle to security is found repeatedly in public discourse. For instance, another *Washington Post* article explicitly argues that encryption hampers law enforcement and offers an explanation of why strong encryption is so dangerous, invoking the threat that it would make it impossible to prosecute paedophiles: '"Apple will become the phone of choice for the pedophile," said John J. Escalante, chief of detectives for Chicago's police department. "The average pedophile at this point is probably thinking, 'I've got to get an Apple phone.'"' (Timberg, 2014). Such references to the threat posed by paedophiles and exacerbated by encryption recur in the discourse. A similar example is the story at the start of this chapter of the rapist who was convicted only because the authorities were able to gain access to his encrypted phone. The depiction of such threats prevails especially in US discourse, although it should be noted that most crimes do not rely on encryption, and state actors have problems citing specific occasions when encryption has actually prevented law enforcement. Nevertheless, identifying paedophiles as the ultimate threat raises the stakes, and very few commentators have been willing to minimise the danger they pose to society.

In general, the debate in Germany focuses less on national security and more on protecting the economy. In 2015, Thomas de Maizière, who was Federal Minister of the Interior at the time, gave a speech in which he depicted encryption as a security threat. This is very much an exception in a German discourse that rarely employs the stark security rhetoric that is prevalent in the United States. Yet, in an echo of his US colleagues, de Maizière framed encryption as primarily a question of balancing privacy and security. He also emphasised the multiple threats arising from networked technology (de Maizière 2015).

End-to-end encryption encodes communications in such a way that even tech companies – the service providers themselves – cannot access them. Decryption is possible only for the user at the other end – the message's designated recipient. When encryption is strong, access by a third party becomes very difficult. Thus, data may remain inaccessible even if a search warrant has been granted. However, while only the user possesses the ability to decrypt, this does not mean that law enforcement might not have other means to access the data. Nevertheless, although the companies focus on presenting themselves as merely acting in the best interests of their customers, this lack of control over access to data is characterised as a hindrance to law enforcement. For instance, in 2015, James Comey made a link between new, end-to-end encryption applications and the threat that these products pose, identifying the mechanism that was designed to regain the trust of customers as a major impediment to FBI investigations:

> In recent months, however, we have on a new scale seen mainstream *products and services designed in a way that gives users sole control over access to their data* [...] For example, many communications services now encrypt certain communications by default, with the key necessary to decrypt the communications solely in the hands of the end user [*sic*] [...] If the communications provider is served with a warrant seeking those communications, the provider cannot provide the data because it has designed the technology such that it cannot be accessed by any third party. When changes in technology hinder law enforcement's ability to exercise investigative tools and follow critical leads, *we may not be able to identify and stop terrorists who are using social media to recruit, plan, and execute an attack in our country.*
>
> (Comey, 2015; emphasis added)

Here, Comey brings together all of the main themes. Encryption – and especially the end-to-end encryption that is increasingly provided by tech companies – puts all of the power over data into the hands of the end user. This argument that law enforcement cannot gain access to data and communication is encapsulated in the term 'going dark'. This refers to the idea that the state cannot gain access to data and thus loses its ability to monitor suspects, and it has become a dominant theme in the discourse. However, it is highly problematic since it evokes a stark dark/light binary that implies there is either full access or no access to data (Schulze, 2017). In reality, there are multiple ways to access data and circumvent encryption (Kerr and Schneier, 2017).[4] Yet, the 'going dark' metaphor advances the idea that society is threatened when the user – and only the user – has control over her data. The citizen is cast as a potential criminal and cause of insecurity, and strong encryption is the tool that empowers him or her. If all communication is not at least technically accessible by the state, so the argument goes, national security is imperilled. Paedophiles, terrorists and murderers are used to depict the possible consequences of strong encryption. With potential criminals using sophisticated technology, strong encryption becomes a hazard. Since every citizen has the potential to become a criminal, having sole control over one's data is a threat.

However, opposition to this line of argument and scepticism about the NSA and FBI have gained ground in the wake of Edward Snowden's revelations. Users are now more concerned about their privacy, and there is more unease about foreign and domestic state agencies' surveillance activities (Hintz and Dencik, 2016; Kuehn, 2018). However, the wave of public outrage has not yet translated into stronger policies aimed at mitigating the influence of the security services (Steiger *et al.*, 2017). Quite the contrary: in 2017 moves were made to increase surveillance by pre-emptively circumventing encryption, but once again these met with fierce criticism in the media (e.g. Spehr, 2017); and in some German states, such as Bremen, attempts to implement new surveillance tools were abandoned (Bröckling, 2018). This shows that the narrative of the state losing control and 'going dark' did not become hegemonic, at least in Germany. Press coverage and public opinion were rather more ambiguous in the UK (see Branum and Charteris-Black, 2015). Meanwhile, in the United States, scepticism about mass surveillance faced a different strategy: a key argument is the insistence that encryption prevents even *legal* access to data. State actors know that it is difficult to prove that access to digital communication is necessary. Activists insist that democracy is threatened by any invasion of privacy, so supporters of weak encryption need to counter this idea. They do this by insisting that strong encryption would prevent access to data even after a warrant had been granted. This theme is consistently emphasised to distinguish legally obtained warrants from illegal mass surveillance. For example:

> We are not asking to expand the government's surveillance authority, but rather we are asking to ensure that we can continue to obtain electronic information and evidence pursuant to the legal authority that Congress provided to us to keep America safe.
>
> (Hess, 2016)

Here, Amy Hess stresses that regulating encryption is not about expanding the rights of the government; it is simply about keeping up to date with the evolving technology (see Comey, 2015 for a very similar argument). Obtaining data is possible if encryption algorithms are cracked or the end device is accessed. Furthermore, access can be facilitated by so-called backdoors, but these are generally considered to increase overall insecurity. Any encryption system can be designed with a backdoor – for instance, if an employer wants to access employees' communications. As I describe in more detail in Chapter 4, there is a long history of the US government trying to implement their own backdoors and forcing companies to insert them in their encryption systems, with the aim of allowing access to private data if a legal warrant permits it. Therefore, according to Hess, backdoors do not substantially decrease privacy, since the state gains access to data only if legally empowered to do so on a case-by-case basis. Thus, the argument is that strong encryption in the hands of the user would merely empower criminals. These 'bad guys' would be able to communicate freely as the government agencies would have no power to intercept their messages. Hence, to stay in control and uphold law and order, the state must retain the power to compel companies to insert backdoors in their encryption systems.

In summary, arguments advocating for weaker encryption are based on the idea that strong encryption hinders law enforcement. In this first narrative, a user who has sole control over her data is considered a threat.

However, since state officials face a legitimacy crisis in the wake of the Snowden revelations, it has proved difficult for this argument to gain ground. This is why they have attempted to establish a distinction between *illegal* mass surveillance and *legal* data access. This narrative relies on the convincing depiction of specific threats that are facilitated by strong encryption in order to counter public objections.

The narrative of the state in control

The arguments outlined above can be summarised as forming a narrative of the state fearing a loss of control. The premise is that the state loses the ability to identify both actual and potential criminals and thus the ability to safeguard innocent citizens and the economy. This narrative rests on the notion that the population is the core site through which security works (Foucault, 2007: 65). Without engaging in the debate over whether Foucault's panopticon can still serve as an appropriate metaphor for conceptualising surveillance (Elmer, 2012), we should note that the state in control narrative is based on the idea of a state knowing about its population. The logic of not only collecting metadata but also having access to the data of the whole (global) population mirrors what Foucault termed the 'function of security':

> The function of security is to rely on details that are not valued as good or evil in themselves, [...] which are not considered to be pertinent in themselves, in order to obtain something that is considered to be pertinent in itself because [it is] situated at the level of population.
>
> (Foucault, 2007: 45)

The 'encryption as a threat' narrative is based on the idea that a user who has sole access to her own data may pose a threat, since she will be able to operate beyond the reach of law enforcement and thus the state. Even a legally obtained search warrant (and legality is an important theme here) will not allow agencies to access this data. Encryption is also acknowledged as a crucial economic tool and part of a digital infrastructure that will strengthen the economy. Yet, it becomes a threat if it enables individuals to become too powerful in relation to the state. The threats posed by terrorists and paedophiles are used to drive home the message that strong encryption can be extremely harmful. Companies and initiatives that allow every user to encrypt their data and hold the only key are thus presented as the central problem for security. However, it is impossible to articulate this narrative of threats without acknowledging the problem of illegitimate mass surveillance and privacy concerns. Public concern over invasions of privacy increased after the Snowden revelations, forcing state actors to react. Nevertheless, although they acknowledge such concerns, they still present too much privacy as a threat to the state's ability to exercise control.

The controversy around encryption depicts it as essential for security, privacy and the economy. Hence, it is not encryption itself that is cast as the threat, but companies' development and distribution of strong encryption software, which supposedly hinders the maintenance of law and order. Ultimately, the debate is about who should have control over and access to citizens' data and communications.

The state out of control

Encryption is not only presented as a threat to national security; it is also discussed as a tool for increasing security and thus for resisting state surveillance, particularly by the NSA. From this perspective, mass surveillance by the state is considered a problem to which encryption is the solution. This position may be condensed in a second narrative that I call 'the citizen as user'. As will become clear, in this narrative the citizen fears state surveillance, and countermeasures focus solely on empowering users as citizens to protect their rights to privacy and freedom.

In the narrative the state is presented as the main threat to the citizen, but companies are also considered problematic to the extent that they cooperate with the state or conduct their own surveillance. Activists view encryption as the most promising way of resisting surveillance. Resistance thus revolves around raising public awareness of the benefits of using strong encryption. However, since the technology is highly complex, citizens must first be 'activated' – that is, educated and put into a position where they feel responsible for their own security. Therefore, in this narrative, security is not provided by the state; rather, it is something to which each individual user must attend.

The state as a threat

Whereas in the first narrative the state presents encryption as a threat and citizens as potential criminals, the counter-narrative portrays the state itself as the main source of insecurity. In this section I will take a closer look at exactly how state agencies and secret services are characterised as threats to understand how users themselves come to be the principal implementers of security measures.

In Germany, debates about encryption and surveillance primarily focus on the NSA as well as the German state and its own secret services, whereas economic espionage is of secondary concern. In both the United States and Germany, a distinction is rarely made between the government and the secret services. In Germany, the NSA is cast as the main threat, but often 'the secret services' in general are depicted as conducting mass surveillance. While other threats, such as criminals and terrorists, are often discussed within a single text in the United States, in Germany surveillance is usually singled out as the main problem, especially in light of the fact that secret services operate outside of parliamentary control. This 'uncontrolled' nature of surveillance, be it carried out by German or US actors, has become a matter of particular concern (Neumann, 2014; Kurz and Rieger, 2013).

In 2014, a German newspaper article even claimed that 'The NSA wants to have all web users under surveillance by 2016' (Mandau, 2014b). After Snowden's revelations, public awareness of surveillance by secret services rose greatly. The topic of internet security came to encompass countering not only economic espionage but also the attentions of the secret services. Indeed, this theme became predominant and in Germany led to the establishment of a parliamentary 'task force' on internet security. Linus Neumann – a prominent activist – concluded that 'the role of secret services, intelligence agencies and their commercial partners needs to be reconsidered on a fundamental level, since they were and still are working to weaken IT security structurally' (Neumann, 2014)

Extra-legal surveillance by secret services around the world and their cooperation with global companies are presented as significant threats. Initiatives by the German state to increase surveillance through the use of trojans ('Staatstrojaner') to collect data have been discussed quite critically. Here, again, we encounter the notion of the state acting contrary to the interests of its own citizens (e.g. Tanriverdi, 2018). Meanwhile, in the United States, the NSA is considered to be winning 'its long-running secret war on encryption, using supercomputers, technical trickery, court orders and behind-the-scenes persuasion to undermine the major tools protecting privacy of everyday communications' (Perlroth *et al.*, 2013). In this *New York Times* article, the theme of secrecy and hidden actions shapes how the issue of internet security, privacy and encryption is discussed. The familiar themes of uncertainty and unknown knowns are also evident in the text. Note how the notion of the citizen as threat has been reversed: now, the state and its secret services are characterised as the principal threat because they indiscriminately target US citizens. The German hacker organisation Chaos Computer Club (CCC) captures this sentiment when stating: 'Rather than investing millions in digital armaments directed *against their own population*, the CCC demands that this money be invested in better technological education' (Chaos Computer Club, 2015; emphasis added). This idea of illegitimate and dangerous state actions forms the cornerstone of this narrative. The danger stems from the state, and the target is the average citizen. In the next section, I will discuss how resistance is imagined and how this paves the way for the individualisation of security.

'The end of all secrets'?

Countering the state and resisting the encroaching surveillance of the secret services is considered a very difficult task, as their power is so great. Indeed, some privacy campaigners lament that the 'war on encryption' has already been lost. The cooperation between private companies, secret services and governments around the world is presented as especially threatening and difficult to overcome. For instance, in a German newspaper article from January 2014, the author discusses the immense power of the NSA and suggests that this will culminate in the creation of a 'quantum computer' that will herald 'the end of all secrets' (Kreye, 2014). According to the author, the 'all-encompassing access of secret services' shows that 'freedom [is] impossible in such a net' (Kreye, 2014). The NSA is frequently characterised as being 'out of control' (Kurz and Rieger, 2013) and impossible to constrain through conventional political means. Indeed, while strategies of resistance are discussed, in general the secret services are presented as unstoppable. The end result is a feeling of disempowerment and resignation among citizens on a global scale (Hintz and Dencik, 2016: 10; Kuehn, 2018: 417).

There have been occasional exceptions to the prevailing sense of powerlessness and the preference for technological solutions – for instance, when the struggle between the FBI and Apple over the decryption of iPhones was debated. In this case, at least the *possibility* of resistance (by Apple) was discussed, with both technological aspects and questions about the roles of private (Apple) and public (FBI) actors in the provision of security featuring in the debate. Mass media devoted hundreds of articles to this battle. Apple tried to make the issue one of public concern and attempted to present itself as a noble corporation that was resisting excessive state power. Activists were more sceptical of the company's motives, but also tried to raise public awareness.

Although activists and experts alike viewed the Apple–FBI stand-off as a political battle, political resistance is often perceived as ineffective. The debate therefore tends to revolve around technology – primarily encryption – as the only hope for resistance against the secret services. Pascal Kuschildgen (2014), for example, presented this view explicitly in a statement to a German parliamentary hearing in 2014. While he demanded new laws that would allow for greater control of the secret services and more internet security, he also asserted that it was unreasonable to suppose that the secret services themselves would allow – let alone offer – such safeguards. In a rather fatalistic tone he concluded: 'Protection from the secret services is not possible, in my opinion. The tool of education [*Aufklärung*] and transparency should be chosen in order to prevent higher damage or interference with the needs of users who need to be protected' (Kuschildgen, 2014). Expecting companies to provide the necessary technological means to resist the secret services is problematic because, historically, they have often cooperated closely with state agencies. Nevertheless, ever since Snowden's disclosures in 2013, several major companies have been cast – and have cast themselves – as potential allies in the fight for privacy (see the extract from Timberg *et al.*, 2013, above). While openly denying any future cooperation with the secret services (or even denying any such cooperation in the past), these companies have insisted that they can and will offer better security through encryption.

Unfortunately, there is a major problem with this technological 'solution': encryption is very difficult to use. Experts and the mass media alike have been keen to stress this point. For instance, writing in the *Washington Post*, Tim Berners-Lee (the developer of the worldwide web) declared:

> Software capable of withstanding NSA snooping is widely available, but hardly anyone uses it. Instead, we use Gmail, Skype, Facebook, AOL Instant Messenger and other applications whose data is reportedly accessible through PRISM. That's not a coincidence: Adding strong encryption to most popular Internet products would make them less useful, less profitable and less fun.
>
> (Berners-Lee, 2013)

Therefore, higher cost and lower usability are significant obstacles in the way of wider adoption of better security systems. Consider this extract from another US newspaper article:

> What's the solution? Make your email more like a letter inside an envelope. The best way to do this is with a process known as encryption, which scrambles a message into unreadable code that needs a key to be unlocked, providing a layer of protection if someone intercepts your email. The downside to encryption tools is that *they are usually difficult to install and use.*
>
> (Wood, 2014; emphasis added)

Wood discusses various programs that are supposedly easy to use and will thus allow more security for every user, although she concludes that face-to-face communication is still the safest form of all. This idea that users are currently unable to use the most appropriate technology for their privacy needs is also evident in discussions of 'one-click encryption', which revolve around how to make programs sufficiently simple that everyone, not only the technological avant-garde, will be able to utilise them. Here, the focus is firmly on company responsibility. Corporations are asked to provide easy-to-use software. From this perspective the user who is unable to use the technology appropriately is the core problem. The solution is then easier-to-use technology and education of the user.

To summarise, technical and political solutions are entangled within this controversy. It has been observed that political solutions are deemed unpromising for preventing surveillance. Meanwhile, technical solutions face the complication that companies might still cooperate with the secret services. Moreover, the encryption by default that is offered by some companies cannot be trusted. The end result is that the individual user who does not know how to encrypt and is unable to utilise the more secure options that are currently available chooses convenience over privacy. This theme will be examined in more detail in the next section.

The activation of the user

The dominant strategy for obtaining and enabling security against state intrusion lies in the activation of the user. Here, activation means focusing on educating and empowering the individual but also shifting responsibility to the level of the individual. The ideas outlined above regarding risk, the power of the secret services and faith in the power of technology have led to this individualisation of security. In the early 2000s, David Lyon suggested that this, in turn, fostered surveillance practices (Lyon, 2003: 21). Here, it is useful to turn his argument around and consider the possibility that surveillance practices may have fostered the individualisation of security. Insecurities are located at the local level, since privacy intrusions can happen to anyone who uses a social network, a webcam or even a mobile phone. Companies might offer encryption by default, but not all do, and they cannot be trusted as they may be cooperating with the secret services. The state is seen as almost omnipotent, and its secret services, by acting covertly and with uncertain accountability, threaten democratic structures. This means they are difficult to resist through legal and political means. Hence, technology is the only way to counter surveillance and data retention. In this context, activation of the user becomes essential.

These technological solutions focus on the actions of every individual user of networked technology. The proper use and application of technology constitutes the core theme when security measures are debated. As a first example, consider the following extract from a newspaper article dealing with an individual who has fallen victim to ransomware:

> So what can we all do to protect ourselves? Keep our computers backed up on an independent drive or by using a cloud backup service like Carbonise, take those software updates and 'patch' alerts seriously and, most of all, Beware the Attachment.
>
> (Simone, 2015)

Encryption is considered to be the best technological tool against surveillance, but again this is viable only by means of 'the correct application of cryptography' (Fachgruppe für Angewandte Kryptographie in der Gesellschaft für Informatik, 2013). Experts highlight that encryption is a complex tool and that securing one's own devices is perceived as a difficult task. Users of these technologies need to understand the basic principles of certain software, how to use it and how to assess it. As a result, suggestions are offered with the aim of informing users about potential security threats, making them aware of the importance of encryption and anonymisation tools, and teaching them how to apply them. The implications of and warnings about technology 'need to be read and understood in order to achieve a higher level of security [...] i.e. consumers have to be sensitised' (Schröder, 2014). Security and comfort are widely presented as two goals that can never be fulfilled at the same time:

The best and most reliable means for defence is giving up comfort. A higher level of security always means less usability of an application. This means that users can only protect themselves by abandoning a specific technology that would make life easier.

(Schröder, 2014)

In this testimony in a German parliamentary hearing, Schröder suggests that greater security is only possible if users become more knowledgeable; otherwise, they will continue to lack the technological literacy they need to understand how to prevent surveillance. Thus, a crucial aim is to educate users so that they understand the security advantages of using complicated encryption software that makes certain activities, such as searching through email messages, more time-consuming. The argument that users need to be educated is also evident in the many attempts by civil society groups to spread knowledge about internet security and encryption as tools for protection (ACLU, 2015; Chaos Computer Club, 2015). Moreover, the demand for better education forms part of governmental discourse (Stevens, 2016: 167–177). This preoccupation with education resembles what Barry identifies as one dimension of the technological society – namely, its concern with 'technical skills, capacities and knowledge of the individual citizen' (Barry, 2001: 3, ch. 6). The main point here is that, for the people to use the technology correctly, they need the necessary knowledge and skills to assess how they should use that technology. In order to participate fully in free democratic practice, everyone needs to become a knowledgeable user. This is a recurring theme in the debate over complex encryption software.

Because of this constellation – the state as a threat, internet security as a ubiquitous risk and technology as the best means of protection – security lies firmly in the hands of the users of technology. This is best exemplified in an interview with a lawyer and an activist in the specialist magazine *Wired* (*Wired*, 2015). The interviewees display great scepticism about the state's actions and suggest that it is always interested in gaining more power. Hence, citizens must attend to their own security: '"I believe that technology holds the necessary answers. We can protect ourselves with its help [...] Encryption becomes always easier, soon everybody can use it. The [David] Camerons of the world can yell as much as they want"' (*Wired*, 2015). Encryption is a powerful tool, but only if it is appropriately applied. Experts from Germany's foremost hacking organisation therefore call for 'digital self-defence' (Chaos Computer Club, 2015). Because the ordinary user is considered to be insufficiently knowledgeable, the tech avant-garde must step up to help. Publicness thus emerges around a socio-technical issue in which the state is the core threat. Hackers can help ordinary citizens by developing programs and offering education. It is thus necessary to listen to these experts and learn how security can be established and improved. Interestingly, the experts frequently ask the state to assist in these activation measures, even though the state is also seen as the principal threat. The state therefore assumes a double role:

chief threat but also one of the means for better security. In addition to educating citizens, it is asked to provide better infrastructure. This should lead to more knowledgeable users who can assess security threats and are able to keep up to date with the rapidly evolving technology. However, the state is not responsible for its citizens' security. Its sole responsibility lies in providing the *conditions* for better security. Ultimately, it is the individual who must act in her own best interests.

This strategy of security measures is encapsulated in the term 'activation'. Users are responsible for their own security, so they need to be educated in order to behave with appropriate judgement. The risk therefore lies at the individual level. This is also the case if threats targeting institutions or companies are discussed. In such instances, each institution is responsible for educating its employees to take the necessary precautions. The logic remains the same: security is increased only if the necessary actions are taken by well-educated users. Therefore, both the proper software and the knowledge to use it are essential.

To summarise the second narrative: the fundamental rights of privacy and freedom are threatened by both state and corporate actions. Thus, internet security is not considered to affect only the private realm. Rather, it is framed as a societal and political issue. Nevertheless, this political problem will be solved only through individual empowerment. Therefore, we may say that creating a citizenry of knowledgeable users is a prerequisite for safeguarding the civil rights of freedom (privacy) and security. Security thereby becomes individualised. It is not provided by the state; instead, it is something to which every individual must attend.

Conclusion: privacy and publicness

This chapter has investigated the encryption security controversy by presenting it in the form of two narratives. In the first, the state is characterised as paranoid and threatened by the user/citizen who uses strong encryption. These user/citizens are empowered by companies and present a risk when they have sole access to their own data. In the second narrative, the state is cast as the main threat because it threatens rather than guarantees the security of its own citizens. In this scenario the citizen needs to be activated, as only an empowered citizen/user can take care of her security. As a result, an individualisation of security is evident. The focus is on how users can change their own behaviour, because the secret services are perceived as so powerful that they cannot be stopped. Some commentators still emphasise the need for political and institutional change and the importance of democracy, but even when they challenge the role of the secret services (e.g. Steinschaden, 2014b), they continue to understand security as an individual responsibility. This dynamic complicates the formation of a form of publicness. Therefore, the 'state out of control' narrative cannot exercise its full force (I shall return to this topic in Chapter 7.).

The analysis in this chapter also brought to the fore several points about controversies and publicness. The public/private dichotomy that might be useful for structuring and describing certain conflicts does not really hold up here. The role of economic actors is especially ambiguous and cannot be neatly located within either of the two main security narratives. This is because such actors both cooperate with and challenge the state. Indeed, the understanding of security shifts from becoming a 'public' good provided by the state to a 'private' good provided by companies or individuals. Distinguishing between public and private is thus of little use from an analytical perspective and indeed on the level of practice. For example, claims to privacy are not necessarily made in opposition to security claims; indeed, they often fit neatly into security narratives. Although state actors need to acknowledge privacy concerns, ultimately it is clear that, for them, security always trumps privacy. Activists challenge this narrative, but even they discuss privacy in a security context. For instance, Karina Rider argues that privacy should not be understood as a value that stands in opposition to state power. Instead, she suggests that demands for stronger encryption and privacy may be read as being 'about matching neo-liberal and law-and-order principles' where the underlying force is determining how 'informational privacy in the market [may] be assured so that American technology firms could dominate world markets' (Rider, 2018: 1383). This speaks to the encryption discourse in which privacy becomes a major aspect of a security discourse. The right to privacy is not espoused to counter surveillance practices; rather, it is becoming part of a broader security discourse (see Mueller, 2010: 160). Experts such as Bruce Schneier argue that security is futile without privacy and thereby make the latter part of the former. In any case, privacy is not seen in opposition to security measures, but as something that we all need in order to remain secure (for the more general debate, see: Bennett, 2011a, 2011b; Gilliom, 2011; Stalder, 2009).

The individualisation of security as it is enacted here does not challenge the state–security nexus, as such. However, it does alter the relationship between state and citizen by framing security as the responsibility of the (private) user rather than the (public) state. This means that 'the public' is not an audience that reacts to the state – in marked contrast to the way in which the audience in securitisation theory is often characterised (Balzacq, 2005). In addition to the ambiguous character of the economy, which provides a moving target, the individualisation of security leads to much more diffuse encryption controversies. They cannot be easily summarised around a for/against binary but rather lead to a diverse set of controversies. What is contested is not just the role of encryption in general security but also more specific questions, such as how encryption could and should be applied and who should be in charge of it.

This also means that one can hardly speak of the emergence of one 'public'. Using the public as a yardstick for assessing encryption controversies would not do justice to the diffuse way in which the role of encryption for security is contested. Publicness emerges around the question concerning the state's and companies' roles in providing security. The responsibility of both the state and businesses becomes a significant issue in technological controversies. The next chapter develops this theme by looking more closely at some of these controversies. The focus will be on technological debates that occur within specialist circles and I will show how prevailing assumptions about encryption technology make such controversies possible. The theme of individualisation of security will reappear when analysing a controversy over who should be able to break certain encryption applications.

Notes

1 To the best of my knowledge, all of the examples the FBI has used to defend its policies have been contested, usually because there is little evidence that encryption actually hindered its investigations.
2 All translations are my own.
3 This quote shows that uncertainty about security threats is frequently linked to the theme of risk. Better encryption will not provide full security; it will merely 'complicate' surveillance.
4 See the discussion on the darknet in Chapter 4.

Bibliography

Primary sources

ACLU (2015). Submission to the Special Rapporteur on the Promotion and Protection of the Right to Freedom of Opinion and Expression, 10 February. Available at: www. aclu.org/aclu-submission-special-rapporteur-encryption-and-anonymity (accessed 23 August 2016).

Balser, M., Gammelin, C., Martin-Jung, H. and Tanriverdi, H. (2017). 'Wanna Cry' Rasender Wurm. *Süddeutsche Zeitung*, 14 May.

Berners-Lee, T. (2013). The high cost of encryption. *Washington Post*, 16 June.

Chaos Computer Club (2015). CCC fordert Ausstieg aus unverschlüsselter Kommunikation. Available at: www.ccc.de/de/updates/2015/ccc-fordert-ausstieg-aus-unvers chlusselter-kommunikation (accessed 20 June 2017).

Comey, J.B. (2015). Director Federal Bureau of Investigation joint statement with Deputy Attorney General Sally Quillian Yates before the Senate Judiciary Committee Washington, DC, 8 July. Available at: www.fbi.gov/news/testimony/going-dark-encryption-tech nology-and-the-balances-between-public-safety-and-privacy (accessed 1 August 2016).

de Maizière, T. (2015). Rede des Bundesinnenmisters beim Forum International de la Cybersécurité. Available at: www.bmi.bund.de/SharedDocs/Reden/DE/2015/01/interna tionales-forum-fuer-cybersicherheit.html (accessed 2 April 2019).

Editorial Board. (2015). The encryption tangle. *Washington Post*, 20 November.

Fachgruppe für Angewandte Kryptographie in der Gesellschaft für Informatik (2013). Kryptographie schützt Grundrechte- gerade im Zeitalter der massenhaften Ausforschung des Datenverkehrs im Internet. Available at: http://fg-krypto.gi.de/presse/ nsa-ueberwachung.html (accessed 15 November 2014)

Gaycken, S. (2014). Öffentliches Fachgespräch des Ausschusses Digitale Agenda des Deutschen Bundestages zum Thema 'IT-Sicherherheit', Schriftliche Stellungnahme von Dr. Sandro Gaycken. Available at: www.bundestag.de/bundestag/ausschues se18/a23/anhoerungen/-/281524 (accessed 16 September 2014).

Hess, A. (2016). Deciphering the debate over encryption, statement before the House Committee on Energy and Commerce, Subcommittee on Oversight and Investigation, Washington, DC. Available at: www.fbi.gov/news/testimony/deciphering-the-debate-over-encryption (accessed 23 April 2019).

Johnson, J. (2015). Remarks by Secretary of Homeland Security Jeh Johnson at the RSA Conference 2015. Available at: www.dhs.gov/news/2015/04/21/remarks-secreta ry-homeland-security-jeh-johnson-rsa-conference-2015 (accessed 13 May 2016).

Kaspersky, E. (2012). AUSSENANSICHT; Angriff aus dem Netz; Hoch entwickelte Computerviren zeigen: Viele Länder bereiten sich auf den Cyber-Krieg vor. Die Attacken können jeden treffen, der einen Internetanschluss hat. *Süddeutsche Zeitung*, 12 September.

Kreye, A. (2014). Digitale Freiheit; der kurze Frühling des Internets. *Süddeutsche Zeitung*, 4 January.

Kurz, C. and Rieger, F. (2013). Snowdens Maildienst gibt auf. Die neuen Krypto-Kriege. *Frankfurter Allgemeine Zeitung*, 9 August.

Kuschildgen, P. (2014). Schriftliche Stellungnahme zum Fragenkatalog für das öffentliche Fachgespräch des Ausschusses Digitale Agenda des Deutschen Bundestages zum Thema 'IT-Sicherheit' am Mittwoch, dem 7.Mai 2014. Available at: www.bun destag.de/bundestag/ausschuesse18/a23/anhoerungen/-/281524 (accessed 16 September 2014).

Mandau, M. (2014a). Die abhörsichere Verschlüsselung. *CHIP*, 1 February.

Mandau, M. (2014b). Wege aus der NSA-Überwachung. *CHIP*, 1 March.

Martin-Jung, H. (2012). Geknackt in 13 Minuten, Forscher zweifeln an der Sicherheit von RSA-Schlüsseln. *Süddeutsche Zeitung*, 27 June.

Neumann, L. (2014). Effektive IT-Sicherheit fördern Stellungnahme zur 7. Sitzung des Ausschusses Digitale Agenda des Deutschen Bundestages. Available at: www.bun destag.de/bundestag/ausschuesse18/a23/anhoerungen/-/281524 (accessed 16 September 2014).

Obama, B. (2016). Transcript of Obama's remarks at SXSW. Available at: www.bos tonglobe.com/news/nation/2016/03/11/transcript-obama-remarks-sxsw/6m8IFsnpJh2 k3XWxifHQnJ/story.html (accessed 13 May 2016).

Perlroth, N. and Sanger, D.E. (2017). Hackers use tool taken from NSA in global attack. *New York Times*, 13 May.

Perlroth, N. *et al.* (2013). NSA able to foil basich safeguards of privacy on web. *New York Times*, 6 September.

Read, M. (2017). Trump is President: Now encrypt your email. *New York Times*, 31 March.

Schmidt, F. (2018). Telegram Der Beschützer des freien Internets Telegram-Gründer trotzt der russischen Medienaufsicht – die sperrt Millionen IP-Adressen *Frankfurter Allgemeine Zeitung*, 19 April.

Schneier, B. (2016). The value of encryption. Available at: www.schneier.com/essays/a rchives/2016/04/the_value_of_encrypt.html (accessed 21 June 2016).

Schröder, T. (2014). Stellungnahme von Thorsteh Schröder zum Fragenkatalog für das öffetnliche Fachgespräch des Aussusses Digtale Agenda des Deutschen Bundestages zum Thema 'IT-Sicherherheit' am Mittwoch, dem 7.Mai 2014. Available at: www. bundestag.de/bundestag/ausschuesse18/a23/anhoerungen/-/281524 (accessed 16 September 2014).

Segal, A. and Grigsby, A. (2016). Breaking the encryption deadlock. *Washington Post*, 14 March.

Simone, A. (2015). My mom got hacked. *New York Times*, 4 January.

Spehr, M. (2017). Dem Staatstrojaner auf der Spur. Kampf gegen die Verschlüsselung: So wollen Ermittler Whatsapp & Co. Aushebeln. *Frankfurter Allgemeine Zeitung*, 12 September.

Steinschaden, J. (2014a). Der Snowden Effekt: Verschlüsselung fürs Volk. Available at: www.netzpiloten.de/der-snowden-effekt-verschluesselung-fuers-volk/ (accessed 23 August 2016).

Steinschaden, J. (2014b). NSA-Überwachung: Verschlüsselung alleine wird uns nicht retten. www.netzpiloten.de/nsa-uberwachung-verschlusselung-alleine-wird-uns-nich t-retten/ (accessed 17 November 2014).

Tanriverdi, H. (2018). IT-Sicherheit Netzpolitiker warnen vor Einsatz des Staatstrojaners. *Süddeutsche Zeitung*, 28 January.

Timberg, C. (2014). FBI chief slams Apple, Google over encryption. *Washington Post*, 26 September.

Timberg, C., Gellman, B. and Soltani, A. (2013). Microsoft moves to boost security. *Washington Post*, 27 November.

Timm, T. (2013). How NSA mass surveillance is hurting the US economy. Available at: www.eff.org/de/deeplinks/2013/11/how-nsa-mass-surveillance-hurting-us-economy (accessed 3 April 2019).

Weaver, N. (2017). Last week's global cyberattack was just the beginning: Two internet plagues have just merged: Without action, they can wreak havoc. *Washington Post*, 15 May.

Wired (2015). Wir befinden uns in einem globalen Informationskrieg! Thomas Drake & Jesselyn Radack im Interview. *Wired*, 29 January.

Wood, M. (2014). Easier ways to protect email from unwanted prying eyes. *New York Times*, 17 July.

Secondary sources

Amoore, L. (2011). Data derivatives: On the emergence of a security risk calculus for our times. *Theory, Culture and Society* 28(6): 24–43.

Amoore, L. and De Goede, M. (2005). Governance, risk and dataveillance in the War on Terror. *Crime, Law and Social Change* 43(2–3): 149–173.

Aradau, C. and Van Munster, R. (2007). Governing terrorism through risk: Taking precautions, (un)knowing the future. *European Journal of International Relations* 13(1): 89–115.

Arquilla, J. and Ronfeldt, D. (2001). *Networks and Netwars: The Future of Terror, Crime, and Militancy*. Santa Monica, CA: Rand Corporation.

Balzacq, T. (2005). The Three Faces of Securitization: Political agency, audience and context. *European Journal of International Relations* 11(2): 171–201.

Barry, A. (2001). *Political Machines: Governing a Technological Society*. London: Athlone Press.

Bauman, Z.*et al.* (2014). After Snowden: Rethinking the impact of surveillance. *International Political Sociology* 8(2): 121–144.

Beck, U. (1986). *Risikogesellschaft: Auf dem Weg in eine Andere Moderne*. Frankfurt am Main: Suhrkamp.

Bennett, C.J. (2011a). In defense of privacy: the concept and the regime. *Surveillance and Society* 8(4): 485–496.

Bennett, C.J. (2011b). In further defence of privacy. *Surveillance and Society* 8(4): 513–516.

BKA (2016). BKA – Internetkriminalität. 28 May. Available at: www.bka.de/DE/UnsereAufgaben/Deliktsbereiche/Internetkriminalitaet/internetkriminalitaet_node.html (accessed 28 March 2017).

Branum, J. and Charteris-Black, J. (2015). The Edward Snowden affair: A corpus study of the British press. *Discourse and Communication* 9(2): 199–220.

Bröckling, M. (2018). Nach Kritik: Verschärfung des Polizeigesetzes in Bremen auf Eis gelegt. Available at: https://netzpolitik.org/2018/nach-kritik-verschaerfung-des-polizeigesetzes-in-bremen-auf-eis-gelegt/ (accessed 14 August 2018).

BSI (2017). BSI – Presseinformationen des BSI – BSI aktualisiert Krypto-Richtlinien der Serie TR-02102. 20 March. Available at: www.bsi.bund.de/DE/Presse/Pressemitteilungen/Presse2017/Aktualisierte_Krypto-Richtlinien_TR-02102_20032017.html (accessed 28 March 2017).

Daase, C. and Kessler, O. (2007). Knowns and unknowns in the 'War on Terror': Uncertainty and the political construction of danger. *Security Dialogue* 38(4): 411–434.

Denning, D.E. (2001). The future of cryptography. In: Ludlow, P. (ed.) *Crypto Anarchy, Cyberstates, and Pirate Utopias*. Cambridge, MA: MIT Press, pp. 85–101.

Dunn Cavelty, M. (2007). *Cyber-security and Threat Politics: US Efforts to Secure the Information Age*. CSS Studies in Security and International Relations. Abingdon andNew York: Routledge.

Elmer, G. (2012). Panopticon–discipline–control. In: Ball, K.Haggerty, K.D. and Lyon, D. (eds) *Routledge Handbook of Surveillance Studies*. Abingdon and New York: Routledge, pp. 21–29.

Elsberg, M. (2012). *Blackout: Morgen ist es zu spät*. Munich: Blanvalet.

Flyverbom, M., Deibert, R. and Matten, D. (2017). The governance of digital technology, Big Data, and the internet: New roles and responsibilities for business. *Business and Society*: 58(1): 3–19.

Foucault, M. (2007). *Security, Territory, Population: Lectures at the Collège de France, 1977–1978*. New York: Picador.

Fuchs, C. (2013). Societal and ideological impacts of deep packet inspection internet surveillance. *Information, Communication and Society* 16(8): 1328–1359.

Gilliom, J. (2011). A response to Bennett's 'In Defence of Privacy'. *Surveillance and Society* 8(4): 500–504.

Guzzini, S. (2000). A reconstruction of constructivism in International Relations. *European Journal of International Relations* 6(2): 147–182.

Hintz, A. and Dencik, L. (2016). The politics of surveillance policy: UK regulatory dynamics after Snowden. *Internet Policy Review* 5(3): n.p.

Kerr, O.S. and Schneier, B. (2017). Encryption workarounds. Available at: www.schneier.com/academic/paperfiles/Encryption_Workarounds.pdf (accessed 25 April 2017).

Krahmann, E. (2011). Beck and beyond: Selling security in the world risk society. *Review of International Studies* 37(1): 349–372.

Kuehn, K.M. (2018). Framing mass surveillance: Analyzing New Zealand's media coverage of the early Snowden files. *Journalism* 19(3): 402–419.

Kutter, I. and Rauner, M. (2012). Blackout: 'Das wäre ein Riesenproblem'. *Die Zeit*, 6 December. Available at: www.zeit.de/2012/50/Stromversorgung-Winter-Blackout-Marc-Elsberg-Jochen-Homann (accessed 29 May 2018).

Lange, M. (2012). *Blackout*. Available at: www.deutschlandfunk.de/blackout.740.de. html?dram:article_id=209278 (accessed 3 April 2019).

Lyon, D. (2003). Surveillance as social sorting: Computer codes and mobile bodies. In: Lyon, D. (ed.) *Surveillance as Social Sorting*. London: Routledge, pp. 13–30.

Lyon, D. (2014). Surveillance, Snowden, and Big Data: Capacities, consequences, critique. *Big Data and Society* 1(2). Available at: https://journals.sagepub.com/doi/full/ 10.1177/2053951714541861 (accessed 3 April 2019).

Mueller, M. (2010). *Networks and States: The Global Politics of Internet Governance*. Information Revolution and Global Politics. Cambridge, MA: MIT Press.

Petersen, K.L. (2012). Risk analysis: A field within Security Studies? *European Journal of International Relations* 18(4): 693–717.

Rid, T. (2013). *Cyber War Will Not Take Place*. Oxford: Oxford University Press.

Rider, K. (2018). The privacy paradox: How market privacy facilitates government surveillance. *Information, Communication and Society* 21(10): 1369–1385.

Schulze, M. (2015). Patterns of surveillance legitimization: The German discourse on the NSA scandal. *Surveillance and Society* 13(2): 197–217.

Schulze, M. (2017). Clipper meets Apple vs. FBI: A comparison of the cryptography discourses from 1993 and 2016. *Media and Communication* 5(1): 54–62.

Smith, S. (2004). Singing our world into existence: International Relations theory and September 11 presidential address to the International Studies Association, February 27, 2003, Portland, OR. *International Studies Quarterly* 48(3): 499–515.

Stalder, F. (2009). Privacy is not the antidote to surveillance. *Surveillance and Society* 1 (1): 120–124.

Steiger, S., Schünemann, W.J. and Dimmroth, K. (2017). Outrage without consequences? Post-Snowden discourses and governmental practice in Germany. *Media and Communication* 5(1): 7.

Stevens, T. (2016). *Cyber Security and the Politics of Time*. New York: Cambridge University Press.

Thorsen, E. (2017). Cryptic journalism: News reporting of encryption. *Digital Journalism* 5(3): 299–317.

UNHCR (2015). *Report of the Special Rapporteur on the Promotion and Protection of the Right to Freedom of Opinion and Expression, David Kaye*. A/HRC/29/32. Geneva: UNHCR.

West, S.M. (2017). Data capitalism: Redefining the logics of surveillance and privacy. *Business and Society*, 5 July. Available at: https://journals.sagepub.com/doi/abs/10. 1177/0007650317718185 (accessed 3 April 2019).

Wikipedia (2018). Blackout (Elsberg novel). Available at: https://en.wikipedia.org/w/ index.php?title=Blackout_(Elsberg_novel)&oldid=836094766 (accessed 29 May 2018).

Yadron, D. (2015). Obama sides with Cameron in encryption fight. *Wall Street Journal*, 16 January. Available at: http://blogs.wsj.com/digits/2015/01/16/obama-side s-with-cameron-in-encryption-fight/tab/print/?mg=blogs-wsj&url=http%253A%252 F%252Fblogs.wsj.com%252Fdigits%252F2015%252F01%252F16%252Fobama-side s-with-cameron-in-encryption-fight%252Ftab%252Fprint (accessed 3 February 2015).

6 Contentious technology

The previous chapter explained how encryption occupies a dual role in security debates, as it is understood as a source of both security and insecurity. In this chapter the focus shifts from security to technology, and to a different set of controversies. While the previous chapter examined how encryption has played out in security controversies, this chapter investigates the controversies revolving around technological issues. I will show how these technological controversies open up spaces of contestation. In contrast to previous critical research within Security Studies, which has often highlighted depoliticising effects, this chapter emphasises the possibility of politicisation. I show how arguments about the uncertainty of knowledge about encryption form part of these controversies. Especially in the case of the dispute between Apple and the FBI (see Chapter 4), publicness occurred around technological issues. The controversy around technological issues was also a contestation of who – the state or Apple – should be the prime provider of security. This chapter concludes the empirical study of the book before Chapter 7 addresses the politics of encryption more explicitly.

First, we need to remind ourselves of the concept of 'technological society', taken from the work of Andrew Barry (2001), which I introduced in Chapter 2. A 'technological society' is defined by the way in which technology becomes a means for framing and solving societal problems. Technologisation is the process by which technologies become a crucial aspect of societal debates and the object of controversies. The effect of a technologised discourse is thus an open process: it can politicise through the emergence of controversies or depoliticise by fostering a technocratic logic. Technocracy, as a political logic, implies bureaucratic structures and decision-making based on the use of the most efficient means, not the outcome of deliberation among different parties (Ginty, 2012: 289). Technocracy is thus, by definition, depoliticising (Burnham, 2001; Zürn, 2016).[1] Focusing on technological solutions can conceal the inherently political character of decision-making processes. For example, for Didier Bigo and his colleagues, the core problem is the overwhelming trust that security professionals place in technologies such as biometrics (Bigo *et al.*, 2010: 56; Bonditti, 2007). Professionals 'focus on mainly technological and non-political solutions to threats to security as opposed to political or diplomatic solutions' (Bigo *et al.*, 2007: 30). In this context, Bigo speaks of the 'myth' that European borders can be controlled, and argues that technological solutions are presented in order to corroborate that myth. He highlights how even small, improvised counter-measures, such as 'aluminium paper to baffle sensors', can jam supposedly sophisticated technology (Bigo, 2005: 77). This shows that technological solutions are never straightforward. However, security professionals refuse to acknowledge this, thereby creating a simplified image of technology. Bigo calls this 'hyper-technologisation' (Bigo, 2005: 78), which leads to depoliti-cisation through a focus on technological solutions that are presented as apolitical. Technology accordingly appears as an 'ultrasolution' (Bigo and Guild, 2005: 255). This research has shown that decisions regarding whether and how to use security technologies are not presented as political. Imple-menting them simply becomes part of a bureaucratic procedure. The result is that the deeply political character of these technologies is obscured. Security technology meets a technocratic logic that favours expertise, bureaucratic structures and supposedly neutral (i.e. technological) solutions rather than open political debate. These observations have been corroborated by other findings in IR and CSS.[2] Focusing more directly on digital technologies, Louise Amoore presents a variety of cases in which algorithms have affected the social world (Amoore, 2013). An emphasis on algorithms not only transforms the understanding of knowledge but also masks the fact that behind these algorithms human decision-making is still decisive. Depolitici-sation occurs in discussions about algorithms, diverting attention from the kinds of decisions that were made and who had the power to make them (Amoore, 2013: 165).

A similar argument about depoliticisation has been made in the context of encryption. In the previous chapter I showed that networked security relies on cost–benefit analyses. Claims about resisting surveillance also rely on this logic: using encryption will not prevent surveillance completely, but it will at least increase its cost. This argument is at the core of most discussions about cybersecurity in which security is framed as the outcome of a cost–benefit analysis. For instance, Seda Gürses and colleagues highlight how eco-nomic arguments are used to criticise mass surveillance practices in debates about encryption (Gürses *et al.*, 2016). They argue that

> technical experts and their supporters have used 'effectivity' as the lens through which to evaluate responses to the surveillance programs and to promote privacy technologies as a way to mitigate against mass surveil-lance, for example, 'encryption makes bulk surveillance too expensive'.
> (Gürses *et al.*, 2016: 583)

For these authors, such arguments might make sense in a certain disciplinary environment, but this logic prevents a broader debate about the political implications of encryption (Gürses *et al.*, 2016: 584). Indeed, 'the legal and technical spheres cannot contribute to the deeply political challenge prompted by Snowden's revelations without themselves undergoing a thorough politicization' (Gürses *et al.*, 2016: 587). Hence, they argue for a politicisation of the encryption discourse, which they perceive as depoliticised due to a focus on technical and economic arguments. This argument speaks to the discussion in the previous chapter, where I showed how an economic logic (the risk-based approach to security) fosters an individualisation of security. Research on technology and depoliticisation is important in highlighting the wider discursive logic in which debates about security technology are often embedded. This research has problematised the naive implementation of security technology and has shown that a focus on technological solutions fosters a depoliticised, technocratic logic. I add to these observations by highlighting instances when practitioners and professionals have failed to present technology in a straightforward way.

This means that I will not repeat the trope of depoliticisation, but instead investigate how we can understand encryption as a contentious and political technology even within these technical debates. A bias towards researching processes of depoliticisation is present not only in the context of technology but also in the field of Security Studies more generally (Hegemann and Kahl, 2016a, 2016b). My argument here is that investigating the contentious character of encryption technology is facilitated by looking at the way in which specific technological aspects become the focus of controversies. And, indeed, it becomes clear over the course of this chapter that, for instance, trust in the capabilities of encryption does not automatically lead to technocracy. Hence, argumentative patterns and themes often associated with depoliticisation and technocracy might also allow for the emergence of controversies.

I begin by reconstructing the 'theory of technology' that is evident in the discourse in order to highlight that practitioners' ideas of encryption are not only shaped by a technocratic logic. The importance of human decisions and the awareness of an uncertain future make it possible for encryption controversies to emerge. The next section looks at how controversies about breaking encryption unfold and how the assessment of who can break encryption is deeply political. I discuss the Apple–FBI dispute and controversies around key-length in more detail. Finally, I conclude by returning to the topic of controversies and depoliticisation.

Theorising technology in practice

In this section I will reconstruct the dominant 'theory of technology' that is shared by experts and political actors alike. This is a first step in identifying how technological controversies are not only governed by a technocratic logic. The empirical analysis was concerned with what one might call a 'theory of technology in practice'. This refers to the assumptions experts have about encryption and how they and mass media present encryption to the public. This approach is based on the insight that theories do not exist only in the academic realm; people in general also avail themselves of theories to impart order to their world. Everybody has assumptions about how the world works, and these everyday theories can be reconstructed by researchers (Schütz, 1932). Members of the general public make assumptions about how technology works or how it will develop (Callon and Latour, 1981; Webster, 2016; Wyatt, 2008). Here, I am most interested in the kind of 'theory' experts have about encryption. How do they describe it? How do they think it is related to other issues? Previous work on technology has similarly reconstructed how experts hold ideas about technological determinism (Peoples, 2010). This is politically significant because these assumptions shape the kinds of political interventions that are considered to be appropriate. I will focus on the themes of uncertainty and complexity and the role of expert knowledge. This will enable me to show that these dominant assumptions open up the way for contesting encryption. From fine-grained analysis of the material described in Chapter 2, I was able to reconstruct the assumptions experts hold on encryption. Quotations will be presented as illustrative examples; the full list of primary sources can be found in the Appendix.

On uncertainty, humans and side-effects

I will use Gary T. Marx's (2007) article 'Rocky Bottoms' as a foil for the present discussion of the dominant 'theory of technology'. In that article, Marx critically discusses prevailing assumptions about technology in a security context. He presents in an ideal-typical fashion beliefs – or 'techno-fallacies' – that shape popular belief systems about technology. According to Marx, these techno-fallacies are inaccurate or oversimplified, but they nonetheless shape public perception and by extension policy-making. I use Marx's ideal-typical reconstruction to show that even though the fallacies he describes might at times be discernible, experts often have a more complicated theory of technology.

One of the techno-fallacies identified by Marx is 'the fallacy of the 100% fail-safe system' in which references to the potency of technologies are treated 'as equivalent to a law of nature' (Marx, 2007: 99). This fallacy is fundamental for a technocratic logic. A logic that is based solely on applying technologies can gain ground only if technology is perceived as omnipotent. People usually have high trust in technology and its capabilities, and no one assumes any faults in the design. This is part of the technocratic effect of technology: since overwhelming trust is placed in technological solutions, experts are empowered and there is no debate over the viability of technological solutions. However, when reconstructing the ideas that shape the perception of encryption, it becomes apparent that, contrary to Marx's fallacy, the debate is *not* primarily shaped by a belief in a powerful technology that never fails. More precisely, digital encryption is considered to be a means of increasing security, but also a source for more insecurity (see Chapter 5). Moreover, its capabilities are deeply contested.

As discussed in the previous chapter, in the context of encryption, security is understood in terms of uncertainty and risk. Experts assume that pre-emptive measures are the best way to improve security, although they are well aware that total security is impossible. The theme of uncertainty also prevails when looking at how knowledge production around encryption is discussed in the more technologically focused arguments. According to Niko Härting's (2014) statement on IT security in Germany, the structure of the internet itself is insecure, and completely secure communication is never possible. From this, it follows that it cannot be attained by improving some aspects, reporting some bugs and so forth. Internet security expert Sandro Gaycken, speaking at the same event, shares Härting's reservations and points out that 'systems are complex [and] IT systems produce effects that [...] cannot be anticipated'. He assumes that IT systems can no longer be controlled or their properties mea-sured (Gaycken, 2014). Part of the problem is that an attack can happen at any time and target any kind of object. This stress on the possibility of failing technology and uncertainty about its behaviour suggest that encryption is not only ruled by a technocratic logic. It is not considered to be an all-powerful tool because human interaction with it is seen as a necessity. In this respect, the encryption discourse differs from Marx's characterisation. Technology is not presented as some kind of black box as the need for human action and unintended consequences are significant parts of the discourse. Contrary to Marx's assertion, experts do not assume a '100% fail-safe system'. This idea of uncertain 'behaviour' by technology is evident in the headline of an article written by the cybersecurity expert Eugene Kaspersky. He states that cyberattacks 'can hit anyone who has access to the internet', and continues:

> Such cyberattacks can sabotage pivotal infrastructure – water reservoirs, air traffic control or the food chain – and have catastrophic effects. Every modern infrastructure is networked to a very high degree [...] Even the attacking country can become the victim of its own weapon; it is called a boomerang effect.
>
> (Kaspersky, 2012)

Here, the insecurity of the technology lies in this 'boomerang effect'. Tech-nology is something that cannot be easily controlled because there is always the possibility that it might fail or act in an unpredictable way.

Later in the same text, Kaspersky (2012) talks about the 'side-effects' of tech-nology. This aspect is dominant when discussing cyberthreats, for instance in an article in which ransomware is described as a 'plague [that] crept up on us' (Weaver, 2017) and can barely be controlled because it involves so many complex factors. This observation speaks to the threat narratives described in the previous chapter: threats become imminent through their description as being ever pre-sent and barely controllable. The side-effects are characteristics of the networked technology that is complex and multi-layered and thus cannot be fully secured (see Chapter 5). What is needed is constant adaptation, for which human input is crucial. This theme runs through the whole debate: human action is necessary because the technology is extremely complex and its effects unpredictable. This means, contrary to a technocratic logic, that technology does *not* determine the solutions for problems. Experts occupy a rather differentiated position in which they acknowledge the need for human action and political decisions. This can be observed in the logic of activation (see Chapter 5). According to this logic, security relies on the education and empowerment of citizens, so more knowledge and awareness on the part of users will improve security. Human action is needed on several levels here: end users need to be knowledgeable; security professionals need to adapt and improve technology; and, although this theme is less present in the discourse, politicians need to provide a legal and institutional framework for activation.

Encryption experts go one step further when they openly admit that they themselves do not know the extent to which encryption can provide security. In a 2013 blog post, the encryption expert Bruce Schneier discusses the likelihood of the NSA breaking encryption systems (Schneier, 2013). Although he trusts the mathematics of encryption, he thinks that poor implementation and end-user behaviour leave it vulnerable. Hence, he feels that expert knowledge is often of limited value. And it is not only experts who have arrived at this conclusion; mass media sources have made similar assessments. For example, in a specialist IT magazine, Uli Ries (2014) writes: 'Even crypto-experts do not know how long current algorithms will provide protection [...] Therefore, it is necessary to inform oneself about possibilities of exchange or upgrade before one [starts] using an encryption solution.' Thus, it is evident that the topic of uncertainty discussed in the previous chapter also appears in the prevailing depiction of expert knowledge. This can also be seen in quotations later in this chapter, which highlight how the limits of knowledge are acknowledged within standard-setting institutions. The value of knowledge for assessing the future is limited. Technology is clearly not just a tool that can be easily applied, and expert knowledge is not an easy fix for complex problems. This notion stands in contrast to a technocratic logic. It opens up space for thinking about the role of human action and, with the emphasis on the limited knowledge of experts, reveals that even resorting to specialists is not a straightforward option.[3] Socio-technical controversies can emerge. In order for encryption to work properly, better technological solutions but also appropriate human actions and adaptations are both needed. These articulations of uncertainty and its side-effects go far beyond the fallacies identified by Marx: that is, misplaced faith in technology to guide decisions and the denial of unintended consequences (Marx, 2007: 99–101).

My analysis revealed that experts do not present technological fixes as the only – or even a straightforward – solution. This becomes especially apparent when analysing texts that look more closely at how encryption works, particularly its dependence on mathematics. The following extract is drawn from *CHIP*, a German IT magazine that deals specifically with encryption and internet security. Mathematics serves as an indicator for the potency of encryption, but at the same time the article problematises the technology:

> According to Snowden, strong encryption offers the best tool to protect oneself against global espionage, but in the way browsers and servers use HTTPS connections, the strength of the encryption does not come into play [...] Furthermore, in PFS the session key is not sent over the internet but computed by every party – this is based on *pure maths*. If PFS is *implemented correctly*, both session keys are deleted as soon as the communication ends.
>
> (Mandau, 2014; emphasis added)

PFS refers to 'perfect forward secrecy', a kind of encryption that works with so-called session keys, which offer better security as they are valid only once. The emphasis on the strengths of mathematics and the strong capabilities of encryption contrasts with its dependency on proper implementation by humans (users) and a technological environment that needs to work as well. This is a theme that runs through the whole debate. Although encryption and its mathematical base are imbued with strong ability to increase security, actors are aware that encryption always relies on implementation by human users and that experts must adapt the technology constantly.

From a cryptographer's perspective, Phillip Rogaway (2015) discusses the 'moral character of cryptographic work'. He problematises many cryptographers' technological optimism and the argument that encryption is just an apolitical branch of mathematics. For Rogaway (2015: 47), 'cryptography is a tool for shifting power' and encryption is inevitably a deeply political technology. His article is an example of an expert's attempt to start a controversy around encryption technology. This political stance of cryptographers also comes to the fore in a report written by a group of esteemed cryptographers (Abelson *et al.*, 2015). This presents a set of technical arguments against most recent government attempts to influence digital encryption. Nevertheless, as the opening of the report makes clear, the authors explicitly situate themselves within contemporary debates on surveillance (Abelson *et al.*, 2015: 5). The discussion of encryption occurs against the backdrop of ongoing societal debates about surveillance – and experts are well aware of this.

In contrast to Marx's analysis of the ever-present perception of technology as unproblematic, omnipotent and neutral, in this section I have demonstrated that many experts present encryption and related technology as highly complex. Not even the knowledge of experts is seen as a straightforward solution, as its limits are acknowledged in the discourse. The tension between trust in encryption and awareness of the difficulties of proper implementation is a dominant theme. Experts frequently acknowledge the limits and the inherent ambiguity of technological fixes. Cryptographers as well as politically active experts and activists present technology in a way that encourages contestation. Even if experts debate 'pure' technology, they are aware of the complex context in which that technology is implemented, and that the effects of encryption depend on its implementation (e.g. Gregg, 2016; Gaycken, 2014; Schröder, 2014). One might say that technological experts have a more 'political' understanding of technology than the professionals whom Bigo and Marx describe. While politicians and security professionals might describe technological solutions in an apolitical way, experts are often aware of the complexity of problems, the influence of uncertainty and the importance of human input.

The power of encryption

We have seen that the open-ended, uncertain character of encryption technology that is heavily dependent on human input is a prevailing theme in the encryption discourse. I have argued that these discursive features allow for a technologisation of the discourse that does not immediately lead to a dominant technocratic logic. However, the discourse is not only shaped by acknowledging uncertainty and ambiguity. Encryption becomes an important issue only because activists, experts and politicians all agree that it is quite powerful.

Indeed, activists and civil society actors highlight the huge capabilities of encryption. For them, the relevance of encryption lies not only in its importance to international finance or the economy in general, but mostly in its ability to safeguard privacy. The Crypto Group of the German Society for Computer Science expresses this in the headline: 'Cryptography protects basic rights – especially in this era of massive espionage of data traffic on the internet' (Fachgruppe für Angewandte Kryptographie in der Gesellschaft für Informatik, 2013). The group characterises encryption as a very powerful tool for protecting privacy. This idea of assigning high capabilities to technology by default is part of a technocratic logic: the power of technology negates the need for human intervention as a technological solution will suffice. According to this technocratic logic, powerful technology overrides the need for political decision-making. However, in the case of encryption it becomes clear that assigning encryption such capabilities is a *precondition* for its politicisation. Encryption is relevant to the struggle between freedom and security only because activists see it as such a powerful weapon in the fight for privacy. Writing in a German newspaper, Andrian Kreye (2014) claims that encryption protects human dignity in the digital realm: 'Cryptography is already much more than a technology for people with industrial or other kinds of secrets. It protects a human right that has not existed for very long: the digital dignity of humans.' Assigning such a capability to technology places it squarely in the political realm. Moreover, the capabilities of encryption are openly acknowledged by state actors. US President Barack Obama, for example, did so in a speech to tech industry leaders when describing the possible disasters that might ensue due to weak encryption:

> [W]e also want really strong encryption, because part of us preventing terrorism, or preventing people from disrupting the financial system or

our air traffic control system or a whole other set of systems that are increasingly digitalized, is that hackers, state or non-state, can just get in there and mess them up.

(Obama, 2016)

Encryption is ascribed power because it provides security from hacking attacks and is crucial for a well-functioning economy. Both activists and politicians therefore see it as a crucial part of the current security landscape. But this idea of a potent technology still does not fit into a technocratic logic, since it does not so much preclude political action as make it essential. Indeed, Obama continued his speech by acknowledging activists' reservations about state interference with encryption systems – 'this is technically true, but I think it can be overstated' – before shifting the question into the social realm and highlighting how this technological question requires political answers: 'And so the question now becomes, we as a society [...] we're going to have to make some decisions about how do we balance these respective risks' (Obama, 2016).

These political questions emerge only because encryption is such a potent tool. This, in turn, demonstrates that characterising encryption as an overarching, ever-present, depoliticising discourse is not a fair reflection of reality. There is more to debates over security technology than Marx's 'techno-fallacies', and encryption does not feature in the controversies as a simple tool that will produce just one outcome: security. As Ries (2014) points out, even experts do not have the solution for complete security. The acknowledged complexity of encryption technology makes it possible to contest its features and disagree about the ways in which it might increase security.

Technocracy rules?

It is too simplistic to assume that discourses in which technology is the focus are always linked to a technocratic logic and its attendant depoliticisation. This would imply the belief that technology is capable of solving problems independently from (political) decisions about its use. Such an understanding is based on the assumption that technology always enters the discourse, and that it is perceived as a neutral tool that will solve problems in a straightforward manner. However, the empirical reconstruction of the understanding of technology showed that the encryption debate is much more nuanced than that. Participants in the debate acknowledge uncertainty about technology, the limits of expert knowledge and the fact that encryption, although a powerful technology, does not provide an easy technological fix for internet security. Therefore, ambiguity is rife when debating encryption, which allows it to inhabit multiple positions within the discourse and opens up space for contestation.

As we saw at the start of this chapter, a great deal of research has shown the technocratic effects of a technologised discourse. In this empirical section, through a reconstruction of the dominant theory of technology, I have shown that contestation is facilitated by the distinct assumptions experts hold about encryption. This allows for controversies to emerge. However, these dynamics are identified only by looking closely at how technology is presented. Only then can we see that specific features of encryption are contentious. In the next section, I will explore this issue further and scrutinise controversies concerning key strength and how to assess the future of encryption security.

Breaking encryption

This section builds on the finding that experts hold assumptions about encryption that do not lead to an apolitical, technocratic logic but instead make controversies possible. However, the analysis shifts to a different site and looks in more detail at how technology becomes part of a controversy. Specifically, I will examine the controversy over the strength of encryption. I will look at various sites – not just governmental institutions – and include multiple actors to highlight the different ways in which the political character of encryption may come to the fore. This is intended as a complement to the previous section, which dealt with the theory of technology in more general terms. The focus on how to break encryption shows, from a different angle, how the political and the technological are intertwined.

One of the most spectacular controversies in this respect was that between the FBI and Apple, which unfolded in spring 2016 (Apple Inc., 2016; Barrett, 2016; Bonneau, 2016; Trail of Bits, 2016). This dispute revolved around the FBI asking Apple to hand over the key to decrypt one particular version of its iPhone (see Chapter 4), which would allow it to gain access to the data of one of the perpetrators of the terrorist attack in San Bernardino, California, in December 2015. Apple refused to hand over the key on the grounds that doing so would grant the FBI access to every other identical device. While similar lawsuits were fought around the same time, Apple's public refusal to comply with the FBI's demands was particularly widely reported and debated around the world. The FBI eventually succeeded in decrypting the phone without Apple's help after an unknown actor intervened and cracked the encyrption, which meant the lawsuit could be dropped. However, before this happened, the question of whether decryption through a third party was actually and legally possible was a matter of significant contestation. Supporters of Apple and the FBI disagreed over whether and how quickly it would be possible to decrypt the phone. However, this argument over a technical issue was entwined with a political struggle over the value of privacy and security.

The most contentious issue related to who would be able to decrypt the phone and what technical knowledge and tools they would need for the task. Even the implications of Apple handing over the decryption key were uncertain: would the FBI gain access to all iPhones, just the suspect's model, or merely a single device? Commentators also disagreed on the question of how easy it would be for Apple to comply with the FBI's demand, whether the FBI had the capability to access the data in another way, and whether they might be able to crack the code themselves. At the one extreme were experts who claimed that only Apple could decrypt the phone, and even then the costs would be immense. At the other, it was categorically stated that US state agencies as well as other actors, including Chinese criminals, had the capability to decrypt iPhones. In 2018 the Department of Justice published a report in which it stated that the FBI had not tried every available technological solution. Instead, it had used the case as a 'poster child' (Department of Justice, Office of the Inspector General, 2018: 8) in order to demand a political solution, thereby supporting the view of Apple and privacy activists. However, it is important to note that before the FBI managed to decrypt the phone with the help of a third party, observers were divided over whether this was possible. These differences point to the uncertainties concerning the capabilities of technology and the fact that it becomes difficult to disentangle the technological and political aspects of such contestations.

This controversy provided two important insights: first, it is only one example of numerous controversies that revolve around what counts as legitimate knowledge; second, it is an example of the contestation of fundamental political categories. With respect to the latter, the controversy was ostensibly about technological questions – namely, who is able to decrypt a particular type of iPhone. But Apple also contested the FBI's role in providing security.[4] By claiming that their customers' interest in privacy – understood as part of security – needed to be prioritised over the FBI's demands, Apple also contested the FBI's understanding of security. It is then not something primarily provided by the state but something companies might be better able to provide to citizens and customers. Moreover, Apple contested the FBI's claims regarding what counts as 'security'. From the company's perspective, keeping data private was more important than having the ability to gain access to knowledge about a terrorist. In this way the controversy speaks to the diverging security narratives identified in the previous chapter. Encryption can be seen either as a tool for hindering national security or as one for strengthening individual security in the wake of increasing invasions of privacy by the state.

Second, this dispute shows that controversies about technology can revolve around what counts as a legitimate claim to knowledge. It forms part of a larger debate about 'going dark' (see Chapters 4 and 5). For years, the FBI has claimed that many encrypted devices hamper its investigations. Yet, it was recently revealed that it has consistently overestimated the number of devices that are encrypted in such a way as to prevent access. Hence, its claim of almost 8000 devices per year had to be corrected in spring 2018. While the Electronic Frontier Foundation (EFF) and *Wired* magazine claimed to have been working on the assumption that the FBI had overstated the numbers all along, the FBI itself blamed the error on faults in its methodology (Barret, 2018; Crocker, 2018; Newman, 2018). Thus, the debate shifted away from technological features and towards procedures and specifically *methods* of knowledge production. A political conflict was conducted through a debate about the limits of technology and the procedures, methods and legitimacy of knowledge production (see also Aradau, 2017). How the FBI arrives at its figures is again not a purely technical question, but a deeply political one. This speaks to the observation, discussed in Chapter 2, that controversies often revolve around the question of what is considered to be reliable evidence (Barry, 2013: 146–147). The question of how often encryption has a negative impact on law enforcement is not easily settled, precisely because so many actors disagree about the evidence and the procedures on which the generation of evidence relies.

One might assume that references to the technological features of encryption could provide straightforward answers to the question of what kind of security it provides. This would depoliticise claims to security via encryption, since references to specific technological features would end the debate. However, the Apple–FBI case shows that technological details can become contentious. Assumptions underpinning public debates about encryption differ starkly. Indeed, a closer look at the seemingly apolitical, technocratic world of bureaucracy shows that encryption is contested on every level. A technological society creates a 'space of government' (Barry 2001: 2) in which technology is regulated and becomes an issue for international and national governance efforts (Jasanoff, 2010). Encryption has traditionally functioned as a weapon, and thus has always been of concern to governments. As it becomes a tool for economic actors and central to the daily lives of citizens, national as well as international institutions increasingly seek to regulate it.

Later, I will look at how encryption is presented in the recommendations issued by these agencies. First, though, it is helpful to remind ourselves of some of the basic technological features of encryption. Encryption technology is based on mathematics – specifically, 'unsolvable' problems – though they are unsolvable only in the sense that no one has yet found a solution. For instance, as we saw earlier, while it is fairly easy to multiply two prime numbers, finding the two prime factors that make a number is much more difficult, so factoring prime numbers is a fundamental aspect of encryption. This has implications for the perceived validity of encryption. The 'security' of a system is not established by absolute proof. Since these 'unsolvable' mathematical problems are the basis of the algorithms that comprise every encryption system, each system relies not on the proof of anything but rather on the hope that no one will ever solve the problem. Higher security simply means that a larger number of experts have tried and failed to crack the algorithm. Apart from solving the mathematical problems, there are several other ways to gain access to encrypted information. Compromising the end device (e.g. a laptop or phone) is one example. For instance, if an end device is compromised, encrypting email will be no use, because information can be read on the device either before it is sent or after it is received. Another way is to get the key. While encryption keys are invariably very long, applications use a much shorter key to unlock the encryption key. For instance, the user does not need to remember the entire encryption key but only a four-digit password when encrypting a device. If someone else gets hold of this password, they will be able to access all of the encrypted information. Another method is the so-called brute-force attack. This simply tries every possible key until the right one is found. It is possible to establish unequivocally how much computing capacity (i.e. brute force) is needed for every possible key to be tried, or so it seems. Key length is crucial in preventing brute-force attacks, especially when securing data against future intrusions. Encrypting or decrypting of messages always requires a key. On average, an attacker will have to try half of all the possible keys before finding the right one. Therefore, a system is easier or harder to break depending on the total number of possible keys and how quickly they can be tried.[5]

Earlier DES encryption systems are now considered too weak and too easily broken (see Chapter 4). And in the future greater computational power will make it possible to break encryption systems that are regarded as secure today. Thus, stronger keys are always needed. This question of future security might seem to be merely a technological one. However, when looking at the discussion of key strength among experts, it soon becomes clear that it is far from easy to settle these debates on wholly technical terms, because various social concerns form part of the argument whenever encryption becomes an issue. The most obvious of these, especially since the Snowden revelations, are the capabilities of state agencies, most prominently the NSA. This means that there is high uncertainty about the means that are necessary to prevent surveillance by the NSA. Actors do not agree on the technical tools that are needed to prevent targeted surveillance. Although the NSA collects bulk data and opposed strong encryption during the 1990s, it does advocate the use of encryption. Uncertainty remains about how security will change with the advent of quantum computers, which the new NSA recommendations supposedly cover (Schneier, 2013; Courtois, 2015). The NSA even recommends certain algorithms, called Suite B. Such recommendations are often seen as credible and, especially since they are mandatory for US government agencies, influence global encryption practices. However, there is also a 'Suite A' about which very little is known, as testified by the subject's meagre Wikipedia entry (Wikipedia, 2019). The important point here is that critics still argue over the NSA's abilities and how much trust should be placed in its recommendations (Bonneau, 2015). Various activist groups and security companies hold very different views of the NSA's technological capabilities. It is thus difficult to provide a thorough assessment of the situation. Although some scenarios outlined by hackers may well be exaggerations, it is worth remembering that people who made comparable claims before Edward Snowden's revelations were accused of paranoia. We simply do not know what the NSA's capabilities are, which systems it can hack and what the costs may be. Of course, there is similar uncertainty about other actors, such as the Chinese government. However, the NSA, through its efforts to cast a veil of secrecy over its actions, is discussed more than any other. This hints at important issues concerning secrecy and the role of secret intelligence services in today's societies, and it also allows us to enter the technical debate about key strength.

Standards for all kinds of encryption systems are issued by the American National Institute of Standards and Technology (NIST), the European Union Agency for Network and Information Security (ENISA) and the German Bundesamt für Sicherheit in der Informationstechnik (BSI – Federal Institute for IT Security). Each of these organisations recommends a specific algorithm and specifies the key size in order to achieve a certain level of security. This standard, which reflects what certain experts consider to be secure, then becomes mandatory for the state agencies of the country in question. For instance, the BSI and NIST both recommend a minimum key size of 128 bit (and they also accept 192 and 256) for AES (Advanced Encryption Standard) and a minimum of 2000 bit (BSI) or 2048 bit (NIST) for RSA.[6] Similarly, the NSA (2016) recommends 256 bit for AES and 3072 bit for RSA (for 'top secret' level). This might sound pretty straightforward, but a closer look at the guidelines reveals that the assessment of security is not as clear cut as one might expect. The recommendations are mostly based on (educated) guesses of how computational power – and thus the cost of specific attacks – will develop in the future. Debating key strength always relates to technological development, but this is far from predictable. Hence, the discussion centres on 'unknown knowns'. Of course, the BSI and NIST try to make the best possible assessments, but assessing risk is not the only factor in achieving security; the discussion also acknowledges inherent *uncertainties* in technological development and thus in the future. Preparing for the future is not achieved by calculating risks that presuppose knowing the main actors, their preferences and the state of the world; rather, it involves handling uncertainty and non-knowledge (Kessler and Daase, 2008). The controversy is ultimately about what is not known, so there are countless references to ignorance and uncertainty in this debate (Aradau, 2017: 328). Ignorance, then, is not merely a blank spot; it is one of the main issues of contention.

One of ENISA's reports on algorithms and key size begins as follows:

> Recommending key sizes for long-term use is somewhat of a *hit-and-miss affair*, for a start it *assumes* that the algorithm you are selecting a key size for is not broken in the mean time. So in recommending key sizes for specific application domains we make an *implicit* [original emphasis] assumption that the primitive, scheme or protocol which utilizes this key size is not broken in the near future. All primitives, protocols and schemes marked as suitable for future use in this document we have confidence will remain secure for a significant period of time.
>
> (ENISA, 2013: 3; emphasis added unless otherwise indicated)

ENISA openly states that its technical recommendations are not founded on secure knowledge but on *assumptions* about future technological developments. When discussing the security of asymmetrical encryption systems, it stresses that these rely on mathematical problems that are difficult to solve.

'Unpredictable scientific progress' may threaten the security of symmetrical systems, but asymmetrical encryption is especially considered to be insecure if quantum computing evolves rapidly (BSI, 2016: 29, 30). Particularly when the aim is to prevent attacks by quantum computers, the assessment becomes highly speculative: 'Moreover, asymmetrical encryption procedures that are recommended in this guideline would become insecure if there were considerable progress in quantum computing' (BSI, 2016: 30).

Quantum computing is generally considered one of the main problems for the future of encryption. But experts are deeply divided over when this will be of practical relevance for encryption (e.g. Buchanan and Woodward, 2017; Nordrum, 2016). At the time of writing, NIST was conducting a public competition to develop quantum-resistant encryption systems. Quantum computing will be able to break encryption systems such as RSA, which rely on prime factoring, with relative ease, whereas the likes of AES will be less affected (although longer keys might be needed). When preparing for the future, security experts need to make assumptions about technological development. Hence, even within bureaucracies, assumptions about encryption shape decision-making. Applying encryption always involves thinking about possible threats, adversaries and future developments. The authors of a NIST report on post-quantum encryption acknowledge that the question of 'when a large-scale quantum computer will be built is complicated and contentious' (Chen *et al.*, 2016: 2). However, because many experts believe that quantum computing will be operational in about 20 years' time, and therefore available to break current encryption systems, 'regardless of whether we can estimate the exact time of the arrival of the quantum computing era, we must begin now to prepare our information security systems to be able to resist quantum computing' (Chen *et al.*, 2016: 2)

Actors try to shape technologies to increase future security, and these standards are developed by governmental organisations in cooperation with scientists. But the awareness of the contentious character of many (scientific) assumptions and the need to work with uncertainties indicate that these dynamics cannot be easily subsumed under a technocratic logic. Debating standards, developing guidelines and especially predicting the future of technological development entails guesswork, acknowledging inherent uncertainty and admitting the limits of expert knowledge. Debating technological capabilities, such as how to break or circumvent an encryption system, is always embedded in a certain context and thus has a political dimension. The technological aspects of the discourse do not necessarily imply concealing this political dimension. Quite the contrary, the contestation of encryption *allows for* a political debate and opens up space for contestation. Again, the notion of technocracy as an umbrella principle for technology's role in the public sphere is inadequate. Indeed, there is not one single controversy, but myriad issues that are contested and bring together a variety of actors, including activists, law-enforcement organisations and standard-setting agencies, such as the NIST. As has been previously stated, it is clear that controversies emerge not only within 'the public'; the contentious character of encryption also comes to the fore when looking at more remote places where controversies might emerge.

Conclusion: encryption beyond technocracy

Rather than repeating the well-known trope of technology having depoliticising effects, this chapter has highlighted the possibilities of emerging controversies. I have analysed the assumptions that experts, politicians and activists hold about encryption and how encryption is presented in the media. Together with the previous chapter, this has revealed that the security discourse revolves around technology and that this technology is contested. I do not think that the technocratic logic previously identified by Gürses *et al.* (2016) in the context of encryption dominates every part of the debate. By shifting the perspective, we can see that ideas about security, such as uncertainty, can actually lead to a different kind of politicisation. Encryption is debated mostly in terms of uncertainty about future developments and actors' capabilities. The importance of human input and the limits of expert knowledge both feature prominently in the discourse on encryption. Dominant assumptions about encryption thus facilitate controversy. We find a variety of recommendations that admit to guesswork regarding the degree of security that encryption might provide. This means that, even in debates in which we might expect an overarching technocratic rationality, uncertainty about technology, including its development and capabilities, continues to shape the discourse. And it is only this uncertainty that makes encryption controversies possible. Encryption does not serve as a tool that will solve problems in a neutral and objective way. Rather, the capabilities, design and development of encryption are all contested. When its development is not viewed as driven by technological determinism, arguments concerning values and the role of human input become subject to debate.

Focusing on the Apple–FBI controversy showed that the legitimacy of knowledge production itself can become part of a controversy, but more importantly it demonstrated that fundamental categories are contested. The state – in the form of the FBI – is no longer considered the prime provider of security. Apple claims to be a more legitimate provider of security for its customers and thereby contests both the FBI's role and the primacy of national security over privacy. In the next chapter this issue and several others will be pursued to allow further reflection on the politics enacted in these controversies.

Notes

1 This formulation is not tautological. It merely shows that (de)politicisation can also be understood as a political strategy.
2 For example, Jenny Edkins has shown similar processes in her study of famine relief programmes in Ethiopia and Eritrea (Edkins, 2000: 200). If technology is required

to solve a specific problem or decide between different options, then experts might be the only people who are capable of making this decision. Famines are presented as a mere technological issue – one that can be solved by choosing the right techniques and applying the right mechanisms. This conceals the inherently political character of these famine relief programmes and the fact that famine is a political issue. Decisions are made on ethical and political grounds, but the technocratic discourse prevents engagement with these grounds by presenting the resulting decisions as 'objectively better' than the alternatives. A similar argument is made by James Ferguson, who researched development programmes in Lesotho (Ferguson, 1990). The political character of these programmes is concealed by focusing on technical aspects that are said to further 'development'. At the same time, political effects, such as the empowerment of a centralised government, are covered up (Ferguson, 1990: 255; see also: Ginty, 2012).

3   Activists explicitly link the political dimension with encryption. The German organisation Netzpiloten emphasises that encryption might be used against surveillance, but the real problem is located on a deeper level. Thus, these (technical) activists especially advocate for a political debate (Steinschaden, 2014). The same sentiment can be found in the actions of organisations such as the EFF and ACLU.

4   As was discussed in Chapter 4, one could read Apple's stance as a clever PR exercise. However, its political significance lies in the way it sparked a controversy about encryption, privacy and security that goes far beyond the company's original motivation.

5   This roughly refers to the available computational power. However, the amount of keys that can be tried can be slowed down if the system has a specific mechanism that requires a certain amount of time to elapse between attempts. This is usually not relevant if humans try different passwords on a device, but it can slow down a brute-force attack conducted by machines. The phone in the Apple–FBI case was fitted with such a mechanism, and one of the FBI's demands was that Apple should provide a way to circumvent it.

6   RSA is an algorithm named after its devisers – Rivest, Shamir and Adleman – and it was the first implementation of public-key cryptography. AES is a symmetrical encryption system that is widely used today (see Chapter 4).

Bibliography

Primary sources

Fachgruppe für Angewandte Kryptographie in der Gesellschaft für Informatik (2013). Kryptographie schützt Grundrechte- gerade im Zeitalter der massenhaften Ausforschung des Datenverkehrs im Internet. Available at: http://fg-krypto.gi.de/presse/nsa-ueberwachung.html (accessed 15 November 2014)

Gaycken, S. (2014). Öffentliches Fachgespräch des Ausschusses Digitale Agenda des Deutschen Bundestages zum Thema 'IT-Sicherherheit', Schriftliche Stellungnahme von Dr. Sandro Gaycken. Available at: www.bundestag.de/bundestag/ausschuesse18/a23/anhoerungen/-/281524 (accessed 16 September 2014).

Gregg, A. (2016). For DC-area firms, enryption is big busines. *Washington Post*, 29 February.

Härting, N. (2014). Schriftliche Stellungnahme zum Fragenkatalog für das öffentliche Fachgespräch des Ausschusses Digitale Agenda des Deutschen Bundestages zum Thema IT-Sicherheit am Mittwoch, dem 7 Mai 2014. Available at: www.bundestag.de/bundestag/ausschuesse18/a23/anhoerungen/-/281524 (accessed 16 September 2014).

Kaspersky, E. (2012). AUSSENANSICHT; Angriff aus dem Netz; Hoch entwickelte Computerviren zeigen: Viele Länder bereiten sich auf den Cyber-Krieg vor. Die Attacken können jeden treffen, der einen Internetanschluss hat. *Süddeutsche Zeitung*, 12 September.

Kreye, A. (2014). Digitale Freiheit; der kurze Frühling des Internets. *Süddeutsche Zeitung*, 4 January.

Mandau, M. (2014). Die abhörsichere Verschlüsselung. *CHIP*, 1 February.

Obama, B. (2016). Transcript of Obama's remarks at SXSW. Available at: www.bos tonglobe.com/news/nation/2016/03/11/transcript-obama-remarks-sxsw/6m8IFsnpJh2 k3XWxifHQnJ/story.html (accessed 13 May 2016).

Ries, U. (2014). Die richtige Krypto-Software: Verschlüsselung ist Vertrauenssache. *Computerwoche*, 2 June.

Schneier, B. (2013). How advanced is the NSA's cryptanalysis: And can we resist it? Available at: www.schneier.com/essays/archives/2013/09/how_advanced_is_the.html (accessed 24 August 2016).

Schröder, T. (2014). Stellungnahme von Thorsteh Schröder zum Fragenkatalog für das öffentliche Fachgespräch des Aussusses Digtale Agenda des Deutschen Bundestages zum Thema 'IT-Sicherherheit' am Mittwoch, dem 7 Mai 2014. Available at: www. bundestag.de/bundestag/ausschuesse18/a23/anhoerungen/-/281524 (accessed 16 September 2014).

Steinschaden, J. (2014). NSA-Überwachung: Verschlüsselung alleine wird uns nicht retten. www.netzpiloten.de/nsa-uberwachung-verschlusselung-alleine-wird-uns-nicht-retten/ (accessed 17 November 2014).

Weaver, N. (2017). Last week's global cyberattack was just the beginning: Two internet plagues have just merged: Without action, they can wreak havoc. *Washington Post*, 15 May.

Secondary sources

Abelson, H.*et al.* (2015). Keys under doormats. *Communications of the ACM* 58(10): 24–26.

Amoore, L. (2013). *The Politics of Possibility: Risk and Security beyond Probability.* Durham, NC, and London: Duke University Press.

Apple Inc. (2016). Customer letter. Available at: www.apple.com/customer-letter/ (accessed 24 February 2016).

Aradau, C. (2017). Assembling (non)knowledge: Security, law, and surveillance in a digital world. *International Political Sociology* 4(1): 327–342.

Barret, D. (2018). FBI repeatedly overstated encryption threat figures to Congress, public. *Washington Post*, 22 May. Available at: www.washingtonpost.com/world/nationa l-security/fbi-repeatedly-overstated-encryption-threat-figures-to-congress-public/2018/ 05/22/5b68ae90-5dce-11e8-a4a4-c070ef53f315_story.html (accessed 16 August 2018)

Barrett, B. (2016). Don't be misled: The Apple–FBI fight isn't about privacy vs. security. *Wired*, 24 February. Available at: www.wired.com/2016/02/apple-fbi-priva cy-security/ (accessed 24 February 2016).

Barry, A. (2001). *Political Machines: Governing a Technological Society.* London: Athlone Press.

Barry, A. (2013). *Material Politics: Disputes along the Pipeline.* RGS-IBG Book Series. Chichester: Wiley-Blackwell.

Bigo, D. (2005). Frontier controls in the European Union: Who is in control? In: Bigo, D. and Guild, E. (eds) *Controlling Frontiers: Free Movement into and within Europe.* Farnham: Ashgate, pp. 49–99.

Bigo, D., Bondiiti, P. and Olsson, C. (2010). Mapping the European field of security professionals. In: Bigo, D.*et al.* (eds) *Europe's 21st Century Challenge: Delivering Liberty.* Farnham and Burlington, VT: Ashgate, pp. 49–63.

Bigo, D. and Guild, E. (2005). Policing at a distance: Schengen and visa policies. In: Bigo, D. and Guild, E. (eds) *Controlling Frontiers: Free Movement into and within Europe.* Farnham: Ashgate, pp. 233–263.

Bigo, D.*et al.* (2007). Mapping the field of the EU internal security agencies. In: Bigo, D.*et al.*, *The Field of the EU Internal Security Agencies.* Paris: Centre d'études sur les conflits/l'Harmattan, pp. 5–66.

Bonditti, P. (2007). Biometrics and surveillance. In: Bigo, D.*et al.*, *The Field of the EU Internal Security Agencies.* Paris: Centre d'études sur les conflits/l'Harmattan, pp. 97–114.

Bonneau, J. (2015). What happened with crypto this year? 2015 in review. Available at: www.eff.org/de/deeplinks/2015/12/technical-crypto-2015-review (accessed 31 March 2017).

Bonneau, J. (2016). A technical perspective on the Apple iPhone case. Available at: www.eff.org/deeplinks/2016/02/technical-perspective-apple-iphone-case (accessed 7 April 2016).

BSI (2016). *Kryptographische Verfahren: Empfehlungen und Schlussellängen; BSI-Technische Richtlinie, Version 2016–01.* Available at: www.bsi.bund.de/DE/Publika tionen/TechnischeRichtlinien/tr02102/index_htm.html (accessed 27 April 2016).

Buchanan, W. and Woodward, A. (2017). Will quantum computers be the end of public key encryption? *Journal of Cyber Security Technology* 1(1): 1–22.

Burnham, P. (2001). New Labour and the politics of depoliticisation. *British Journal of Politics and International Relations* 3(2): 127–149.

Callon, M. and Latour, B. (1981). Unscrewing the big Leviathan: How actors macro-structure reality and how sociologists help them to do so. In: *Advances in Social Theory and Methodology: Toward an Integration of Micro- and Macro-Sociologies.* London: Routledge, pp. 277–303.

Chen, L., Jordan, S., Liu, Y.-K., Moody, D., Peralta, R., Perlner, R. and Smith-Tone, D. (2016). *Report on Post-Quantum Cryptography.* National Institute of Standards and Technology. Available at: https://nvlpubs.nist.gov/nistpubs/ir/2016/NIST.IR. 8105.pdf (accessed 25 April 2019).

Courtois, N.T. (2015). NSA plans to retire current cryptography standards. *Journal of Scientific Apartheid*, 15 September. Available at: http://blog.bettercrypto.com/?p=1917 (accessed 31 March 2017).

Crocker, A. (2018). FBI admits it inflated number of supposedly unhackable devices. Available at: www.eff.org/de/deeplinks/2018/05/fbi-admits-it-inflated-number-supp osedly-unhackable-devices (accessed 4 April 2019).

Department of Justice, Office of the Inspector General (2018). *A Special Inquiry Regarding the Accuracy of FBI Statements Concerning Its Capabilities to Exploit an iPhone Seized during the San Bernardino Terror Attack Investigation.* Washington, DC: Oversight and Review Division, US Department of Justice.

Edkins, J. (2000). *Whose Hunger? Concepts of Famine, Practices of Aid.* Borderlines. Minneapolis and London: University of Minnesota Press.

ENISA (2013). *Algorithms, Key Sizes and Parameters Report: 2013 Recommendations.* Available at: www.enisa.europa.eu/publications/algorithms-key-sizes-and-parameters-report (accessed 25 April 2019).

Ferguson, J. (1990). *The Anti-politics Machine: 'Development', Depoliticization, and Bureaucratic Power in Lesotho.* Cambridge: Cambridge University Press.

Ginty, R.M. (2012). Routine peace: Technocracy and peacebuilding. *Cooperation and Conflict* 47(3): 287–308.

Gürses, S., Kundnani, A. and Van Hoboken, J. (2016). Crypto and empire: The contradictions of counter-surveillance advocacy. *Media, Culture and Society* 38(4): 576–590.

Hegemann, H. and Kahl, M. (2016a). (Re-)politisierung der Sicherheit? *Zeitschrift für Internationale Beziehungen* 23(2): 6–41.

Hegemann, H. and Kahl, M. (2016b). Security governance and the limits of depoliticisation: EU policies to protect critical infrastructures and prevent radicalisation. *Journal of International Relations and Development* 21(3): 1–28.

Jasanoff, S. (2010). Ordering knowledge, ordering society. In: Jasanoff, S. (ed.) *States of Knowledge: The Co-production of Science and Social Order.* International Library of Sociology. London: Routledge, pp. 13–45.

Kessler, O. and Daase, C. (2008). From insecurity to uncertainty: Risk and the paradox of security politics. *Alternatives: Global, Local, Political* 33(2): 211–232.

Marx, G.T. (2007). Rocky bottoms: Techno-fallacies of an age of information. *International Political Sociology* 1(1): 83–110.

Newman, L.H. (2018). 'Significant' FBI error reignites data encryption debate. *Wired*, 23 May. Available at: www.wired.com/story/significant-fbi-error-reignites-data-encryption-debate/ (accessed 4 April 2019).

Nordrum, A. (2016). Quantum computer comes closer to cracking RSA encryption. Available at: https://spectrum.ieee.org/tech-talk/computing/hardware/encryptionbusting-quantum-computer-practices-factoring-in-scalable-fiveatom-experiment (accessed 30 December 2018).

NSA (2016). Commercial National Security Algorithm (CNSA) Suite. Available at: https://apps.nsa.gov/iaarchive/customcf/openAttachment.cfm?FilePath=/iad/library/ia-guidance/ia-solutions-for-classified/algorithm-guidance/assets/public/upload/Commercial-National-Security-Algorithm-CNSA-Suite-Factsheet.pdf&WpKes=aF6woL7fQp3dJi6FG9CPPXwjcrp3QgfxLehJh7 (accessed 7 June 2016).

Peoples, C. (2010). *Justifying Ballistic Missile Defence: Technology, Security and Culture.* Cambridge Studies in International Relations. Cambridge and New York: Cambridge University Press.

Rogaway, P. (2015). The moral character of cryptographic work. *IACR Cryptology ePrint Archive* 2015: 1162.

Schütz, A. (1993). *Der Sinnhafte Aufbau der Sozialen Welt: Eine Einleitung in die Verstehende Soziologie.* Frankfurt: Suhrkamp.

Trail of Bits (2016). Apple can comply with the FBI court order. Available at: http://blog.trailofbits.com/ (accessed 24 February 2016).

Webster, M.D. (2016). Examining philosophy of technology using Grounded Theory methods. In: *Forum Qualitative Sozialforschung/Forum: Qualitative Social Research* 17(2). Available at: www.qualitative-research.net/index.php/fqs/article/view/2481/3948 (accessed 4 April 2019).

Wikipedia (2019). NSA Suite A cryptography. Available at: https://en.wikipedia.org/wiki/NSA_Suite_A_Cryptography (accessed 4 April 2019).

Wyatt, S. (2008). Technological determinism is dead: Long live technological determinism. In: Hackett, E.J. (ed.) *The Handbook of Science and Technology Studies.* 3rd edition. Cambridge, MA: MIT Press, pp. 165–180.

Zürn, M. (2016). Opening up Europe: Next steps in politicisation research. *West European Politics* 39(1): 164–182.

7   The politics of publicness

Controversies about the role of encryption for security, its technical capabilities and future development occur in the context of wider debates about security, privacy and surveillance. Importantly, these controversies do not only occur within established political institutions and they are not best analysed with reference to the presence or absence of a public sphere. The term 'crypto-politics' suggests not only that encryption is investigated as a political issue, but also that political practices occur in different places and follow a different logic.

This concluding chapter begins by briefly summarising earlier findings before teasing out how encryption controversies and new forms of publicness enact a certain kind of politics. A core theme of the book has been the political significance of security practices facilitated by networked technology that cross established boundaries of public and private, political and technological. Therefore, the analytical challenge is to avoid 'uncritically fall[ing] back on reiterating familiar institutional repertoires of democratic action' (Huysmans, 2016: 82). This theme of trying to locate political practices outside established institutions motivated the empirical study on encryption controversies. This interest shaped the empirical analysis, and even more so the conceptual vocabulary, developed in Chapters 2 and 3. Using controversies and the notion of publicness allowed me to establish the political significance of encryption controversies and to engage with the question of how they transgress established boundaries. Highlighting these insights is the main theme of this chapter.

In Chapter 2 I outlined my interest in the social study of technology. The aim was to understand particular encryption controversies not as isolated events, but in their wider social context. I identified Andrew Barry's work on material politics as particularly valuable for this project as it opens up the category of the political. In Chapter 3 I developed this topic by engaging with the literature on publicness. The discussion of Callon and colleagues' and Marres's work highlighted the danger of conceptualising as 'truly' political only those controversies that feed back into the political system of governmental decision-making. Drawing on John Dewey, I introduced the notion of publicness, which enables the researcher to grasp multiple and heterogeneous forms of publicness that have the potential to challenge political institutions fundamentally. Therefore, the aim of the analysis was not to assess the absence or presence of a public, but to enquire how controversies emerge around technological and societal questions. The final section of this chapter will return to the question of the kind of politics that is enacted in encryption controversies.

My interest in political practices outside established institutions also influenced my chosen methodology, which alternates between empirical research and theoretical work (see Chapter 2). Drawing on insights from Grounded Theory and pragmatist epistemology more generally, I conducted a textual analysis of debates about encryption. This fine-grained analysis was corroborated by additional empirical research on the wider context in which encryption is embedded. In Chapter 4 I described the technological and historical context of encryption, which enabled a better understanding of the debates that I explored in Chapters 5 and 6. The empirical emphasis of actors such as activists, experts and the media allowed a theoretical interest in new forms of politics to be translated into a methodology for empirical research.

In this book, my interest in the politics of technology is focused firmly on the idea of publicness. As Chapters 4, 5 and 6 testify, forms of publicness appear in a variety of spaces and might not look like the traditional image of a public sphere. For example, hackers and activists engage with society at large with the intention of educating them about encryption. Opposing state initiatives to regulate encryption, these dissidents openly advocate its wider use. Indeed, after the Snowden revelations, activists were vocal in challenging state agencies and their claim that strong encryption generates more insecurity by encouraging users to 'go dark' and thus prevents the state from accessing encrypted digital data. I describe these two narratives in Chapter 5. The contestation of encryption is not limited to the issue of whether it is a source of or a threat to security. Also contested are its technological features and their implications for the future security of the internet. In addition, I discussed the individualisation of security and how strategies of 'activation' that aim to educate the user are prevalent in encryption debates. Privacy, in this context, is also becoming part of this individualised understanding of security. Every user needs to attend to security through self-education and implementation of the appropriate technologies. In Chapter 6, I explored this theme by showing how Apple challenged the state's traditional role as the prime provider of security. Apple claimed to provide better security for its users and contested the FBI's understanding of the issue.

On a more conceptual level, I explained why the categories of public and private, which are routinely used to describe societal struggles and the responsibilities of actors, do not hold much analytical value when trying to understand the contestation of encryption. The qualifiers 'public' and 'private' are used to denote specific realms, but also to describe the structure of societal struggles (Newman, 2007). When it comes to security, the state is usually considered to be the chief provider of security, which in turn is understood as a public good

(Loader and Walker, 2007). However, the struggles around encryption do not follow the conventional division between public and private actors. The role of businesses in particular is much more complex because ICT companies are entangled in encryption controversies in manifold ways. Traditionally, companies have opposed the state's attempts to regulate strong encryption, but they have also cooperated with the state's attempts to conduct mass surveillance. In order for their business models to work, some companies need unencrypted data in order to target advertisements. In addition, the study of encryption and security has shown that the contestation of technology occurs not only within the public sphere in the form of public deliberation, but also among experts and activists. Chapter 6 points out that experts and standard-setting agencies acknowledge the limits of expert knowledge. Future technological development and what counts as legitimate knowledge are both elements in ongoing encryption controversies. Experts are aware that their knowledge is tentative and that encryption will never provide total security. Rather than accepting the dominant technocratic logic, I have explained how technological controversies become aspects of political controversies. What is more, the technological and political elements of these controversies are closely entangled. I have also shown how contestations about values and technology appear as distinct political struggles. In the next section I will explain how these controversies have both reaffirmed and transcended established political categories.

The politics of publicness

Publicness is a concept that highlights how controversies emerge in remote spaces around socio-technical problems, and it can help us to grasp where and how contestation about security technologies occurs. In Chapter 3, I argued that the concept of controversy, as developed by Marres and Callon and colleagues, is always entangled in a language game that takes the nation-state for granted and locates the political within its boundaries. The viewpoint underlying Callon and colleagues' approach is that controversies must be translated through hybrid forums into the realm of governmental decision-making in order to become truly political. While I argued in Chapter 3 that this conceptualisation is problematic for methodological and conceptual reasons, the political ramifications of the approach are now apparent, too. Callon and colleagues' concept of controversies imposes a limit on what may be considered 'politically' relevant.

A similar problem emerges in IR and especially in Security Studies in that what may be classified as 'political' tends to be understood quite narrowly. IR, as a discipline, is constituted through the separation of international and domestic realms as well as high politics and the everyday (Huysmans, 2011: 375; Neocleous, 2006; Walker, 2010).[1] The idea of a sovereign nation-state is at the centre of the discipline and defines the core categories and what counts as a legitimate research object. Indeed, the practice of defining political community through the concept of sovereign states seems entirely natural, while alternative approaches to thinking about political community are rarely, if ever, advanced (Neocleous, 2000: 8; Bartelson, 2009). Most social and political theory simply accepts 'the givenness of orders [as] an object of investigation' (Isin, 2008: 26). However, it is important to bear in mind that nation-states are neither natural nor inevitable (Walker, 1997: 62), and, as has been argued throughout the book, technological controversies might not be captured by an analysis that takes categories such as public and private not as inevitably tied to governmental insitutitons.

The notion of publicness widens our gaze and allows us to engage with the dispersed ways in which controversies emerge. As a result, we can engage with forms of contestation that cannot be meaningfully linked back to governmental institutions. In the empirical analysis, 'the political' was not the defining category. Instead, I entered the analysis of the politics of encryption through a focus on controversies. The aim was to initiate a discussion on politics, technology and security that used a particular vantage point – publicness – for debating the political. Again, my intention was not to locate the democratic deficits in encryption politics and evaluate the degree of 'publicness'. The empirical study started with the concept of publicness as a tool for teasing out how contestation about technology and security occurs in a variety of settings. In the following sections the forms of publicness identified in the previous chapter will be presented again, and I will highlight the particular politics that came to the fore.

Public and private

Throughout this book, I have problematised the analytical value of the categories 'the public' and 'the private' as two realms that are already constituted. In the context of encryption, actors are not easily placed into either category. Indeed, the state is presented as a threat in debates about encryption and surveillance, and it does not emerge as the public provider of security. This observation is corroborated by recognising that the role of the economy is ambivalent. Generally speaking, e-commerce relies on the availability of strong encryption, but some companies depend on unencrypted data traffic for targeted advertising. The role of business cannot be easily described as on the side of either public or private. The struggle between the FBI and Apple can be read as a highly sophisticated PR campaign by the corporation to show its customers how much it values their privacy. However, at the same time, technological issues and more fundamental debates about security and privacy were important aspects of the dispute, giving rise to a kind of politics that cannot be understood as a mere PR stunt. Apple's actions challenged the state's role as the prime provider of security. More generally, I showed that an individualisation of security is dominant in the context of encryption. Private companies are competing with the state to define 'security' and how it may best be established. As a result, the idea of security as a public good is being reconfigured.

The distinction between public and private relies on the concept of a sovereign nation-state having control over its borders in a world system of states. But the boundaries between states are challenged by new practices emerging in relation to networked technology. The problem of reading the politics of encryption into a spatial logic is evident when looking back to when digital encryption became an object of contention in the 1990s. As described in Chapter 4, Phil Zimmermann uploaded an encryption program to the web, thereby making it available to people around the world. He thus 'exported' encryption, which was forbidden at the time, supposedly in the interests of national security. However, the encryption export regulations were very weak because they could not account for networked technology. Hence, it is easy to refute any discourse that rests on the assumption that governing encryption within the boundaries of a nation-state is feasible.[2] The technology contradicts the spatial logic of nation-states: publicness is linked to the activity of 'exporting' (or uploading) encryption, making it available around the world and, in doing so, making claims about the need to protect privacy.

Chapter 3 highlighted how forms of publicness emerge in relation to specific issues. A public is not a pre-existing realm. What is more, publics are not preconditions for issue formation, but its result. This point is crucial for any discussion of security, as it creates some analytical space to think about publics and politics independently of the nation-state, in which the public always has a prescribed role. The concept of the public can be used to describe a sphere of state activities, but the role of the state as the provider of public goods is changing, as I have discussed in the context of security. In addition to the state being perceived as a threat, the citizen as the beneficiary of security is replaced by the user. In Chapter 5 I showed how security measures focus on users and their activation. As Janet Newman (2007) has shown, escaping the familiar (gendered) concept of the public and developing truly novel approaches towards this issue has been difficult; moreover, escaping a national logic often proves impossible. Glimpses of the ways in which publicness is enacted appear once we look beyond state politics. We can then see how Apple and activists challenge the role of the state and even more so the accompanying idea of national security. Importantly, challenging concepts of national security does not work through extending the notion of security, as was done on a theoretical level in CSS. Challenging the idea of security as a public good that is provided by the state works through reconfiguring the relationships between citizen and user, and global corporations and security agencies.

The state and publicness

Especially in the context of digital technology, where the focus often lies on its alleged novelty, it is important to remember the role of state politics and the impact of its regulatory power. Here we can see how political practices occur inside as well as outside established political institutions. Governmental institutions and international organisations are still important sites for understanding encryption as a political technology. Contestations about encryption that took place during the 1990s and 2010s brought into being a kind of publicness that revolved around state politics and standard-setting. The question of regulating the export of encryption has been at the heart of the struggle over encryption, as today are questions of the extent to which policies regulate the implementation of encryption. For example, in India, China and Saudi Arabia the implementation of encryption is strongly regulated, giving the state greater control (see GP Digital, 2018). Claims for the implementation of backdoors and the weakening of encryption are still made in the name of national security. Concepts of national security are therefore prevalent in the encryption discourse. Activists who oppose strong regulations on encryption do so mostly by arguing that weak encryption means weak security and by pointing out that strong encryption is not a significant obstacle to law enforcement. Thus, the narrative of national security and the importance of the state is not challenged. In these instances, state politics remains the decisive feature, and ignoring the importance of the state's ability to set standards risks overemphasising new features of technology and neglecting state power.

Forms of publicness also emerge in governmental institutions. The politics of encryption is enacted in parliamentary debates, the implementation of policies and statements by elected politicians. Throughout the empirical analysis, it was clear that activists, in the form of NGOs, try to pressure the government into the implementation of certain policies. Law-enforcement agencies also try to engage with 'the public' in order to influence policy decisions. This part of the debate could thus be most easily summarised as a binary debate of the state versus civil society. Whereas the state has tried to securitise encryption, activists have tried to show the inherent insecurities of weak encryption. For instance, hackers launched global initiatives to demonstrate the inherent insecurity of the DES algorithm. This is a prime example of the emergence of publicness around a specific algorithm and its technological capabilities. However, while the debate revolved around a technological issue, publicness still followed the basic binary of the state versus civil society. Even though aspects of the politics of encryption cannot be fully grasped within such a framework, these forms of more 'traditional' politics remain relevant.

Security and liberty

The political battle over encryption began with the question of whether modern encryption algorithms could and should be protected as war technologies or whether their 'civilian' purpose – protection against eavesdropping by third parties – justified some relaxation of the regulations. It is possible to read this dispute as one between two competing values of security and liberty. On the other hand, it may also be read as a debate about what 'security' actually means. Security can be understood as a good that is provided by the state, or the state can be seen as a cause of insecurity (Loader and Walker, 2007). While the sovereign state protects its citizens, it is also a source of insecurity (Walker, 1997: 68). Being a citizen is a precondition of being a subject of security, but at the same time the state – the Leviathan – not only guarantees the security of its citizens but also represents a potential threat. For some people, the state is necessary to guarantee basic rights, whereas others consider it a threat to ensuring those rights. Contemporary debates about the role of the state thus 'tend to express diverging judgments about how historical transformations have or have not changed the conditions under which it might be rational to gamble that Hobbes was right' (Walker, 1997: 67). The debates we have observed throughout this book are thus about whether the state is safeguarding or threatening the right to privacy. These dynamics play out in encryption debates, where dissidents and state agencies disagree over the degree to which sovereign power should be tamed and how that power should be assessed vis-à-vis a loss of security. Controversies emerge over the assessment of encryption but also the role of the state. However, as even dissident discourse cannot really escape the hegemonic vocabulary of national security, it is unable to articulate a different way of thinking about security. What is contested is thus the relative power of state agencies and how their claims about security should be balanced with demands for more privacy, not the idea of national security itself.

Chapter 5 ends with a discussion of how concepts of privacy are linked with ideas about democracy and liberty, but ultimately subsumed under security: privacy becomes part of security. From the state's perspective, privacy is appreciated as long as it does not interfere with security; and even dissident voices discuss privacy in terms set by the rhetoric of security. This is one of the reasons why the dissident discourse cannot develop full force. However, it is now possible to see that this apparent inconsistency brings to the fore a problematic tension within the concept of security itself. It is no accident that privacy and, by extension, freedom are framed as parts of security. As Mark Neocleous shows in his reading of Hobbes and Locke, the idea of a liberal state is not based on balancing security and liberty; rather, it always favours security. Locke's concept of prerogative is conceptualised in such a way that the sovereign is endowed with the right to impose extraordinary measures in the name of security (Neocleous, 2007: 134, 141). Contrary to a traditional (liberal) reading of Locke, Neocleous shows how he 'concedes that there must also be scope for discretion, since the public good – the protection of life, liberty and property – may sometimes require immediate action' (Neocleous, 2007: 135). Through a wider reading of liberal authors, Neocleous concludes that liberty is subsumed under security in the liberal tradition (Neocleous, 2007: 141; see also Neocleous, 2000). These dynamics are apparent in the encryption discourse, where claims to privacy are always presented in the language of security. Ultimately, then, privacy is subsumed under security. This discourse is not an exceptional case; rather, the way in which privacy plays out in security discourse reaffirms the liberal values that underlie the debate. The privacy and security debate thus remains squarely within the boundaries of security, as understood in terms of the nation-state. Privacy, as a signifier, cannot unleash its full force to counter the 'state in control' narrative. It remains firmly subordinated to security.

Citizenship

Activists who oppose surveillance make more radical claims when challenging the primacy of national security. Organisations such as the EFF, the ACLU and the CCC oppose the argument that surveillance and weak encryption are necessary for higher security. Note that their dissent focuses on an assessment of the relationship between encryption and security, not on the need for security as such. This dissident discourse questions dominant state practices around the world. As we saw in Chapter 5, these activists present the state as a core threat for their privacy and, by extension, for freedom and democracy. Their critique ranges from moderate demands for increased scrutiny of secret service activity to more radical criticism of the role of security agencies in general. They also demand better education of users and the widespread implementation of strong encryption, especially in popular platforms such as WhatsApp and Skype. However, even this dissident discourse is deeply embedded in assumptions about a world order that consists of nation-states. More radical demands to abolish all of the world's secret services are easily dismissed as utopian and neglectful of the importance of national security. This speaks to Rob Walker's more general observation that dissident voices invariably have great difficulty in countering hegemonic ideas about national security (Walker, 1983). For example, dissidents who criticise the spread of nuclear weapons and advocate for peace cannot escape the image of state sovereignty as the prime organising principle of world politics. Critiques that focus on ethical arguments, and therefore challenge the predominant paradigm of inter-state war, are easily dismissed as naive (Walker, 1983: 316). Activists who propose alternative ideas about security and the role of security agencies are similarly scorned.

The dissident discourse runs into further problems when considering the role of citizenship. The encryption discourse relies on the sovereign state – a fact that is thrown into sharp relief when we consider how citizenship enters the discourse. The US discourse, in particular, continues to be shaped by the idea that citizens and non-citizens are subject to two different surveillance regimes. The collection and analysis of foreigners' data is generally considered as legitimate in the interests of national security, whereas US citizens' data is protected by much stricter rules. The security discourse on encryption and surveillance is thus based on a citizen/non-citizen distinction, in which US citizens enjoy a higher degree of protection. Even dissidents take the distinction between citizens and non-citizens for granted. Meanwhile, in Germany, this distinction is reflected in the distinction between foreign and domestic surveillance. The Snowden revelations spurred a debate about foreign powers' collection of 'German' data and their monitoring of German citizens. The distinction between inside and outside, citizens and non-citizens, runs through the whole discourse. This observation reaffirms a point made in Chapter 3, where I argued against reducing the question of politics of technology to one about better citizen engagement. Within STS scholarship, practices such as hybrid forums are the predominant perspective from which questions about the public and citizens are asked. From this perspective, the question is how to increase citizen engagement in order to improve citizen–state relations and thus increase the legitimacy of political decisions (Irwin, 2001). But this approach again presupposes where 'true' political practices (namely, only those that form part of citizens' engagement practices) might be found, and predefines the legitimate subjects and bearers of rights. These practices favour an ideal of the 'neutral' citizen over engaged activists (Laurent, 2016; Young, 2001). Therefore, emerging publicness is again predefined as linked to citizenship, and again only a very specific kind of practice that occurs within these engagement initiatives is considered to be politically significant. As a result, the notion of citizenship again circumscribes what can be counted as political.

To be clear, there are alternative conceptualisations of citizenship. One of these presents 'acts of citizenship' primarily as claims of rights that challenge the status quo (Isin, 2008; Isin and Ruppert, 2017). Critical research has shown that the category of citizenship is not quite as static as one might think when using the vocabulary of sovereign states (Muller, 2004; Nyers, 2006). Here, the focus has been on how the act of claiming citizenship entails the possibility of enacting a form of politics that does not reaffirm the idea of the sovereign nation-state (Guillaume and Huysmans, 2013). Indeed, one might think that citizenship can be understood only by looking at how it is embedded in global 'regimes of citizenship' (Guillaume, 2014; see also Hindess, 2002). Citizenship, as a category, thus possesses critical potential. However, my point is that this potential is scarcely evident in the discourse on encryption. Rather, citizenship, as it is embedded in this discourse, reaffirms the boundary between inside and outside the state. Even dissident voices cannot escape this constitutive distinction and reaffirm the ideas of national security and citizens as the prime political subjects.

The extent to which the idea of nationality and citizenship frames this debate is somewhat surprising, since the networked and global character of the internet works against the citizen/non-citizen distinction when it comes to regulating encryption. While the regulation of surveillance might be structured around membership of a particular state, in practice it is almost impossible to monitor non-citizens without also implicating citizens. Data travels around the world and is saved on servers in a variety of countries, and communication occurs between citizens and non-citizens. Therefore, no distinction can be made between 'domestic' and 'international' data. This can be observed, for instance, in the controversy over the Foreign Intelligence Surveillance Act (FISA). Even though the aim is to monitor foreigners – particularly (potential) terrorists – in practice this often necessitates the collection and analysis of US citizens' data (Goitein and Patel, 2015). Understandably, this is contested by activists on the grounds that FISA violates the privacy of those citizens. Thus, the citizen/non-citizen distinction is inscribed into the dispute. Activists are well aware of this, but they implicitly reaffirm the underlying logic by insisting that the surveillance of non-citizens must be curbed in order to protect citizens. The protection of non-citizens is therefore only a by-product of the protection of citizens.

However, activists also challenge established political categories. For instance, as shown in Chapters 4 and 5, they challenge the need for secrecy and the legitimacy of the secret services. Attempts to internationalise problems of encryption and surveillance are embedded in a national framework, but a pre-figuration of alternative orders can be observed, too. Publicness has emerged around socio-technical issues that cannot be captured by known institutional repertoires. In contrast to the hybrid forums depicted by Callon and colleagues, these forms of publicness are less geared towards legitimising or prescribing specific policies that are executed by state agencies.

In summary, contestation of encryption generates publicness that is not always directly linked to state policies. From the perspective of classical democratic theory, one would expect forms of publicness to be analysed in terms of the lack of a public sphere that is sufficiently powerful to influence state policies. However, through the notion of publicness, it is possible to identify controversies occurring both inside and outside established institutions and then reflect on the politics enacted in those controversies. The political implications of this kind of publicness develop their force only when they are *not* read back into a framework of national sovereignty and security. As discussed earlier, a public sphere as depicted by liberal democratic theory never truly existed. This is one reason why the notion of publicness developed here is more helpful when engaging with the politics of socio-technical controversies. A focus on the public can reify certain ideas of what may be classified as political and what democratic practices ought to look like (see Dean, 1999, 2001). Accordingly, the notion of publicness should not simply be read back into a framework of national politics in which the purpose of any form of publicness is ultimately to legitimise political decisions.

Conclusion

To conclude, this book has provided an empirical study of digital encryption – a technical issue that has received little attention within the field of political science. The aim was thus twofold: to provide an empirical study of encryption understood as a political technology; and to engage conceptually with publicness as a way to think about political practices beyond the nation-state. In order to achieve this I used insights from the social study of science and technology in order to grasp the political character of encryption controversies. I introduced the notion of publicness to highlight how controversies have the potential to challenge established political categories. This led on to an empirical study of digital encryption in which I showed that this is a deeply contentious technology and that 'political' and 'technological' questions are invariably entangled in disputes over the issue. Studying these controversies thus demands sensitivity to how contestation about encryption as a security technology spans these two dimensions. Reconstructing the two main security narratives, I showed that encryption is presented as both a threat to and a solution for security. This ambiguous characterisation comes to the fore in multi-layered debates over encryption technology that encompass questions about mathematics and the implementation of technology, but also political values. Indeed, these controversies revolve around specific technological features and even what may be considered legitimate evidence. In this final chapter the empirical results have been revisited to show that the focus on publicness not only provides methodological access to the empirical analysis, but also allows for a broadening of the way we think about security and politics. Through the notion of publicness, we can see how the encryption discourse oscillates between reaffirming and challenging established political institutions.

Notes

1 CSS research has tried to address this problem by broadening the concept of security to include notions such as human security. However, these efforts do not dissociate politics from sovereignty (Huysmans, 2006: 38; Walker, 1997: 64), so they do little to help us think about political practices outside established institutions.
2 Similarly, initiatives by activist groups such as Resetthenet focus on the United States and the surveillance activities of the NSA, yet they need to claim that they will alter the internet on a global scale. Combating the (national) threat is possible only when people around the world take action. The contradictions of the security discourse are thus worked into the dissident discourse.

Bibliography

Bartelson, J. (2009). *Visions of World Community.* Cambridge and New York: Cambridge University Press.
Dean, J. (1999). Making (it) public. *Constellations* 6(2): 157–166.
Dean, J. (2001). Publicity's secret. *Political Theory* 29(5): 624–650.
Goitein, E. and Patel, F. (2015). What went wrong with the FISA court? Available at: www.brennancenter.org/publication/what-went-wrong-fisa-court (accessed 6 July 2016).
GP Digital (2018). World map of encryption laws and policies. Available at: www.gp-digital.org/world-map-of-encryption/ (accessed 8 November 2018).
Guillaume, X. (2014). Regimes of citizenship. In: Isin, E. and Nyers, P. (eds) *Routledge Handbook of Global Citizenship Studies.* Abingdon: Routledge, pp. 150–158.
Guillaume, X. and Huysmans, J. (2013). Citizenship and securitizing: Interstitial politics. In: Guillaume, X. and Huysmans, J. (eds) *Citizenship and Security: The Constitution of Political Being.* PRIO New Security Studies. Abingdon: Routledge, pp. 18–34.
Hindess, B. (2002). Neo-liberal citizenship. *Citizenship Studies* 6(2): 127–143.

Huysmans, J. (2006). *The Politics of Insecurity: Fear, Migration, and Asylum in the EU*. The New International Relations. London and New York: Routledge.

Huysmans, J. (2011). What's in an act? On security speech acts and little security nothings. *Security Dialogue* 42(4–5): 371–383.

Huysmans, J. (2016). Democratic curiosity in times of surveillance. *European Journal of International Security* 1(1): 73–93.

Irwin, A. (2001). Constructing the scientific citizen: Science and democracy in the biosciences. *Public Understanding of Science* 10(1): 1–18.

Isin, E. (2008). Theorizing acts of citizenship. In: Isin, E.F. and Nielsen, G.M. (eds) *Acts of Citizenship*. London: Zed Books, pp. 15–43.

Isin, E. and Ruppert, E. (2017). Digital citizenship and surveillance: Citizen Snowden. *International Journal of Communication* 11. Available at: https://ijoc.org/index.php/ijoc/article/view/5642 (accessed 5 April 2019).

Kratochwil, F. (1995). Sovereignty as dominium: Is there a right of humanitarian intervention? In: Lyons, G.M. and Mastanduno, M. (eds) *Beyond Westphalia? State Sovereignty and International Intervention*. Baltimore, MD: Johns Hopkins University Press, pp. 21–42.

Laurent, B. (2016). Political experiments that matter: Ordering democracy from experimental sites. *Social Studies of Science* 46(5): 773–794.

Loader, I. and Walker, N. (2007). *Civilizing Security*. Cambridge: Cambridge University Press.

Muller, B.J. (2004). (Dis)qualified bodies: Securitization, citizenship and 'identity management'. *Citizenship Studies* 8(3): 279–294.

Neocleous, M. (2000). Against security. *Radical Philosophy* 100: 7–15.

Neocleous, M. (2006). From Social to National Security: On the fabrication of economic order. *Security Dialogue* 37(3): 363–384.

Neocleous, M. (2007). Security, liberty and the myth of balance: Towards a critique of security politics. *Contemporary Political Theory* 6(2): 131–149.

Newman, J. (2007). Rethinking 'The Public' in Troubled Times: Unsettling state, nation and the liberal public sphere. *Public Policy and Administration* 22(1): 27–47.

Nyers, P. (2006). The accidental citizen: Acts of sovereignty and (un)making citizenship. *Economy and Society* 35(1): 22–41.

Walker, R.B.J. (1983). Contemporary militarism and the discourse of dissent. *Alternatives: Global, Local, Political* 9(3): 303–322.

Walker, R.B.J. (1997). The subject of security. In: Krause, K. and Williams, M.C. (eds) *Critical Security Studies: Concepts and Cases*. London: UCL Press, pp. 61–82.

Walker, R.B.J. (2010). Democratic theory and the present/absent international. *Ethics and Global Politics* 3(1): 21–36.

Young, I.M. (2001). Activist challenges to deliberative democracy. *Political Theory* 29(5): 670–690.

Appendix

Primary sources

ACLU (2014). The path to privacy reform: After one year of Snowden revelations, what comes next? Available at: https://live-aclu-d7.pantheonsite.io/other/path-priva cy-reform (accessed 20. 06. 2017).

ACLU (2015). Submission to the Special Rapporteur on the Promotion and Protection of the Right to Freedom of Opinion and Expression. 10 February. Available at: www.aclu.org/aclu-submission-special-rapporteur-encryption-and-anonymity (accessed 23 August 2016).

Balser, M., Gammelin, C., Martin-Jung, H. and Tanriverdi, H. (2017). 'Wanna Cry' Rasender Wurm. *Süddeutsche Zeitung*, 14 May.

Brunnermeier, M. (2018). Bitcoin brauchen staatliche Kontrolle. *Frankfurter Allgemeine Zeitung*, 23 March.

Berners-Lee, T. (2013). The high cost of encryption. *Washington Post*, 16 June.

Chaos Computer Club (2015). CCC forder Ausstieg aus unverschlüsselter Kommunikation. Available at: www.ccc.de/de/updates/2015/ccc-fordert-ausstieg-aus-unverschlusselter-kom munikation (accessed 20 June 2017).

Cohn, C. (2014). Nine epic failures of regulating cryptography. Available at: www.eff. org/de/deeplinks/2014/09/nine-epic-failures-regulating-cryptography (accessed 20 June 2014).

Comey, J.B. (2015). Director Federal Bureau of Investigation joint statement with Deputy Attorney General Sally Quillian Yates before the Senate Judiciary Committee Washington, DC, 8 July. Available at: www.fbi.gov/news/testimony/going-dark-encryp tion-technology-and-the-balances-between-public-safety-and-privacy (accessed 1 August 2016).

de Maizière, T. (2015). Rede des Bundesinnenmisters beim Forum International de la Cybersécurité. Available at: www.bmi.bund.de/SharedDocs/Reden/DE/2015/01/ internationales-forum-fuer-cybersicherheit.html (accessed 2 April 2019).

Editorial Board (2015). The encryption tangle. *Washington Post*, 20 November.

Fachgruppe für Angewandte Kryptographie in der Gesellschaft für Informatik (2013). Kryptographie schützt Grundrechte- gerade im Zeitalter der massenhaften Aus-forschung des Datenverkehrs im Internet. Available at: http://fg-krypto.gi.de/presse/ nsa-ueberwachung.html (accessed 15 November 2014)

Farrel, H. (2017). Cybercriminals have just mounted a massive worldwide attack: Here's how NSA secrets helped them. *Washington Post*, 13 May.

Gaycken, S. (2014). Öffentliches Fachgespräch des Ausschusses Digitale Agenda des Deutschen Bundestages zum Thema 'IT-Sicherherheit', Schriftliche Stellungnahme von Dr. Sandro Gaycken. Available at: www.bundestag.de/bundestag/ausschuesse18/a 23/anhoerungen/-/281524 (accessed 16 September 2014).

Gregg, A. (2016). For DC-area firms, enryption is big busines. *Washington Post*, 29 February.

Grehsin, M. (2014). Deutschland hat angefangen. *Frankfurter Allgemeine Zeitung*, 5 January.

Gröger, A.-C. (2016). Einbruch leicht gemacht; Viele Firmen vernachlässigen grundlegende IT-Sicherheitsregeln. Hacker freut das. *Süddeutsche Zeitung*, 28 April.

Gross, G. (2016). Digital rights group: Save security, reject FBI's iphone unlocking request, Available at: www.pcworld.com/article/3044810/security/digital-rights-group -save-security-reject-fbis-iphone-unlocking-request.html (accessed 10 August 2016).

Harlan, C. (2018). Riding the bitcoin wave in search of a golden shore. *Washington Post*, 4 February.

Härting, N. (2014). Schriftliche Stellungnahme zum Fragenkatalog für das öffentliche Fachgespräch des Ausschusses Digitale Agenda des Deutschen Bundestages zum Thema IT-Sicherheit am Mittwoch, dem 7 Mai 2014. Available at: www.bundestag. de/bundestag/ausschuesse18/a23/anhoerungen/-/281524 (accessed 16 September 2014).

Hawkins, D. (2018). The Cybersecurity 202: We surveyed 100 security experts: Almost all said state election systems were vulnerable. *Washington Post*, 21 May.

Henning, P.J. (2018). Policing cryptocurrencies has become a game of whack-a-mole for regulators. *New York Times*, 1 June.

Hess, A. (2015). Statement before the House Oversight and Government Reform Committee, Subcommittee on Information Technology, Washington, DC, 29.04.2015. Available at: www.fbi.gov/news/testimony/encryption-and-cyber-security-for-mobile-el ectronic-communication-devices (accessed 23 August 2016).

Hess, A. (2016). Deciphering the debate over encryption, statement before the House Committee on Energy and Commerce, Subcommittee on Oversight and Investigation, Washington, DC. Available at: www.fbi.gov/news/testimony/deciphering-the-debate-o ver-encryption (accessed 23 April 2019).

Hildebrandt, T. (2015). Psst, wir werden abgehört. *Die Zeit*, 11 June.

Johnson, J. (2015). Remarks by Secretary of Homeland Security Jeh Johnson at the RSA Conference 2015. Available at: www.dhs.gov/news/2015/04/21/remarks-secreta ry-homeland-security-jeh-johnson-rsa-conference-2015 (accessed 13 May 2016).

Kaspersky, E. (2012). AUSSENANSICHT; Angriff aus dem Netz; Hoch entwickelte Computerviren zeigen: Viele Länder bereiten sich auf den Cyber-Krieg vor. Die Attacken können jeden treffen, der einen Internetanschluss hat. *Süddeutsche Zeitung*, 12 September.

Kreye, A. (2014). Digitale Freiheit; der kurze Frühling des Internets. *Süddeutsche Zeitung*, 4 January.

Krüger, P.S. (2016). AUSSENANSICHT; Ein Datenschützer namens Apple; Der Technologiekonzern weigert sich, ein iPhone für das FBI zu knacken. Das dient dem Geschäft – und auch den Bürgern. *Süddeutsche Zeitung*, 16 March.

Kuhn, J. (2014a). Datenschutz: Der Wert des Privaten. *Süddeutsche Zeitung*, 5 June.

Kuhn, J. (2014b). Komplett verschlüsselt; Der Kurznachrichtendienst WhatsApp setzt auf die Privatsphäre seiner Nutzer. *Süddeutsche Zeitung*, 20 November.

Kurz, C. (2017). Algorithmische Daumenschrauben bleiben nicht ohne Wirkung. *Frankfurter Allgemeine Zeitung*, 24 July.

Kurz, C. and Rieger, F. (2013). Snowdens Maildienst gibt auf. Die neuen Krypto-Kriege. *Frankfurter Allgemeine Zeitung*, 9 August.

Kuschildgen, P. (2014). Schriftliche Stellungnahme zum Fragenkatalog für das öffentliche Fachgespräch des Ausschusses Digitale Agenda des Deutschen Bundestages zum Thema 'IT-Sicherheit' am Mittwoch, dem 7.Mai 2014. Available at: www.bun destag.de/bundestag/ausschuesse18/a23/anhoerungen/-/281524 (accessed 16 September 2014).

Mandau, M. (2014a). Die abhörsichere Verschlüsselung. *CHIP*, 1 February.

Mandau, M. (2014b). Wege aus der NSA-Überwachung. *CHIP*, 1 March.

Martin-Jung, H. (2012). Geknackt in 13 Minuten, Forscher zweifeln an der Sicherheit von RSA-Schlüsseln. *Süddeutsche Zeitung*, 27 June.

Martin-Jung, H. (2013). Wehrlose Technik; Selbst Amateure knacken heute Geheimcodes. *Süddeutsche Zeitung*, 23 July.

Mascolo, G. and Richter, N. (2016). Überwachung. Behörden soll Verschlüsselung knacken. *Süddeutsche Zeitung*, 23 June.

McKune, S. (2015). Encryption, anonymity, and the 'right to science'. Available at: www.justsecurity.org/22505/encryption-anonymity-debates-right-science/ (accessed 23 August 2016).

Moorstedt, M. (2014). Die Lüge von den Metadaten; Was die Telefon-Verbindungen, welche die Geheimdienste speichern, wirklich offenbaren. *Süddeutsche Zeitung*, 19 March.

Neumann, L. (2014). Effektive IT-Sicherheit fördern Stellungnahme zur 7. Sitzung des Ausschusses Digitale Agenda des Deutschen Bundestages. Available at: www.bun destag.de/bundestag/ausschuesse18/a23/anhoerungen/-/281524 (accessed 16 September 2014).

Obama, B. (2016). Transcript of Obama's remarks at SXSW. Available at: www.bos tonglobe.com/news/nation/2016/03/11/transcript-obama-remarks-sxsw/6m8IFsnpJh2k 3XWxifHQnJ/story.html (accessed 13 May 2016).

Ochs, S. (2016). Apple's latest legal filing: 'The founders would be appalled'. Available at: www.pcworld.com/article/3045042/security/apples-latest-legal-filing-the-founders-woul d-be-appalled.html (accessed 20 June 2017).

Perlroth, N. and Sanger, D.E. (2015). FBI Director repeats call that ability to read encrypted messages is crucial. *New York Times*, 19 November.

Perlroth, N. and Sanger, D.E. (2017). Hackers use tool taken from NSA in global attack. *New York Times*, 13 May.

Perlroth, N. and Shane, S. (2013). Waging privacy fight as FBI pursued Snowden. *New York Times*, 3 October.

Perlroth, N.*et al.* (2013). NSA able to foil basic safeguards of privacy on web. *New York Times*, 6 September.

Read, M. (2017). Trump is President: Now encrypt your email. *New York Times*, 31 March.

Ries, U. (2014). Die richtige Krypto-Software: Verschlüsselung ist Vertrauenssache. *Computerwoche*, 2 June.

Schmidt, F. (2018). Telegram Der Beschützer des freien Internets Telegram-Gründer trotzt der russischen Medienaufsicht – die sperrt Millionen IP-Adressen *Frankfurter Allgemeine Zeitung*, 19 April.

Schneier, B. (2013). How advanced is the NSA's cryptanalysis: And can we resist it? Available at: www.schneier.com/essays/archives/2013/09/how_advanced_is_the.html (accessed 24 August 2016).

Schneier, B. (2016). The value of encryption. Available at: www.schneier.com/essays/a rchives/2016/04/the_value_of_encrypt.html (accessed 21 June 2016).

Schrader, C. (2013). Daten sollen nicht mehr Ware und Währung sein: Akademie fordert Kultur der Privatheit im Internet. *Süddeutsche Zeitung*, 16 May.

Schröder, T. (2014). Stellungnahme von Thorsten Schröder zum Fragenkatalog für das öffetnliche Fachgespräch des Aussusses Digtale Agenda des Deutschen Bundestages zum Thema 'IT-Sicherherheit' am Mittwoch, dem 7.Mai 2014. Available at: www. bundestag.de/bundestag/ausschuesse18/a23/anhoerungen/-/281524 (accessed 16 September 2014).

Segal, A. and Grigsby, A. (2016). Breaking the encryption deadlock. *Washington Post*, 14 March.

Simone, A. (2015). My mom got hacked. *New York Times*, 4 January.

Soldt, R. (2016). Illegaler Waffenhandel im Netz: Aus dem Darknet in Haft. *Frankfurter Allgemeine Zeitung*, 28 July.

Spehr, M. (2017). Dem Staatstrojaner auf der Spur. Kampf gegen die Verschlüsselung: So wollen Ermittler Whatsapp & Co. aushebeln. *Frankfurter Allgemeine Zeitung*, 12 September.

Steinke, R. (2017). Internet Polizei schanppt Darknet-Größe. *Süddeutsche Zeitung*, 12 June.

Steinschaden, J. (2014a). Der Snowden Effekt: Verschlüsselung fürs Volk. Available at: www.netzpiloten.de/der-snowden-effekt-verschluesselung-fuers-volk/ (accessed 23 August 2016).

Steinschaden, J. (2014b). NSA-Überwachung: Verschlüsselung alleine wird uns nicht retten. www.netzpiloten.de/nsa-uberwachung-verschlusselung-alleine-wird-uns-nicht-retten/ (accessed 17 November 2014).

Tanriverdi, H. (2018). IT-Sicherheit Netzpolitiker warnen vor Einsatz des Staatstrojaners. *Süddeutsche Zeitung*, 28 January.

Timberg, C. (2014). FBI chief slams Apple, Google over encryption. *Washington Post*, 26 September.

Timberg, C., Gellman, B. and Soltani, A. (2013). Microsoft moves to boost security. *Washington Post*, 27 November.

Timm, T. (2013). How NSA mass surveillance is hurting the US economy. Available at: www.eff.org/de/deeplinks/2013/11/how-nsa-mass-surveillance-hurting-us-economy (accessed 3 April 2019).

Weaver, N. (2017). Last week's global cyberattack was just the beginning: Two internet plagues have just merged: Without action, they can wreak havoc. *Washington Post*, 15 May.

Wired (2015). Wir befinden uns in einem globalen Informationskrieg! Thomas Drake & Jesselyn Radack im Interview. *Wired*, 29 January.

Wood, M. (2014). Easier ways to protect email from unwanted prying eyes. *New York Times*, 17 July.

Index

access to digital encryption 3–4
ACLU *see* American Civil Liberties Union
Acting in an Uncertain World 40
activation of user 102–4, 116
activism 3–4, 98–100, 119, 139
Actor-Network Theory 11, 19, 45
acts of citizenship 140
Acuto, Michele 20
add-ons 73
Advanced Encryption Standard 61, 124–5
AES *see* Advanced Encryption Standard
algorithms 60–64, 67, 96, 112, 122, 137–8
alleged novelty of digital technology 10, 23, 46, 136
ambiguity 11, 22, 45, 72, 81, 91–2, 105, 119
American Civil Liberties Union 4, 29, 139
American Civil War 63
American pragmatism 39
Amicelle, Anthony 22
Amoore, Louise 112
analytical leverage 10
analytics of devices 7
anonymisation technologies 4, 60, 70, 102–3
anonymity 59, 66–74; cryptowars I 66–71; cryptowars II 71–3; 'dark net' and TOR 73–4
ANT *see* Actor-Network Theory
Apple 12, 28, 72, 85, 90, 92, 94, 100, 111, 113, 120–22, 126
application flaws 62
assemblage 20
assumptions about future technological developments 124–6

asymmetrical encryption systems 61–2, 124–5
AT&T 4, 68–9
automated border security 21

backdoors 70, 96
banning encryption 68, 89, 94, 103
Barry, Andrew 7, 10, 18, 21–6, 39, 75, 83, 103, 111, 132
Barthe, Yannick 39
Berners-Lee, Tim 101
Best, J. 9
Big Data 7, 21
Bigo, Didier 19, 111–12, 118
biometrics 111–12
Bitcoin 65–6
BKA *see* Bundeskriminalamt
black-boxing 20, 41, 115
Blackout 83–4
blockchains 63–6
blogging 84, 116
boomerang effect 115
Bourdieu, Pierre 19, 24–5
breaking encryption 120–26
brute-force attack 123
BSI *see* Bundesamt für Sicherheit in der Informationstechnik
Bundesamt für Sicherheit in der Informationstechnik 89, 124
Bundeskriminalamt 82
Bush, George W. 71

Caesar algorithm 61
Callon, Michel 11, 39–46, 51–3, 132–4, 141
Cambridge Analytica 5
Cameron, David 68, 94, 103
causality 48
Cavelty, Dunn 83

CCC *see* Chaos Computer Club
CCTV 5, 21
Chaos Computer Club 28–9, 59, 99, 103,
    139
Charlie Hebdo 68, 71
CHIP 85, 116–17
ciphertext 61
citizen–state relations 140
citizenship 10, 44, 137–41
Clipper phones 68–70
codebreaking 63
collective action 46, 48
collective agency 48
Comey, James 88, 94–5
commercialisation of internet security 65
concept of publicness 134–41; citizenship
    139–41; public vs private 135–6;
    security and liberty 137–9; state and
    publicness 136–7
conceptualising encryption 97
conduct of warfare 7
consensus conferences 42–3
constant comparison method 31
contestation of technology 59
controlling encryption 64, 69–70
controlling technology 59
controversies around encryption 31–2
cookies 73
cost–benefit analyses 112
counter-publics 9, 53
cryptanalysis 63
crypto anarchy 1
crypto-politics 132
cryptocurrencies 65–6
CryptoNet 1
cryptowars 64, 66–73; 2.0 71; round 1
    66–71; round 2 71–3 war on
    encryption 99–100
cyber risks 81–6
cyberattack 83–6, 115
cybercrime 82, 89, 95
cyberthreat 83–4, 93
cyberwar 31
'cypherpunks' 59, 75

'darknet' 73–4
data capitalism 92
Data Encryption Standard 61, 64, 69,
    123, 137
datafication 92
de Maizière, Thomas 29, 94
debating encryption 3–4
decryption 58–79, 85, 120–21
delegative democracy 42–3

democratic deficit 135
Denning, Dorothy E. 69–70
Department of Homeland Security 89
Department of Justice 121
depoliticisation 39, 112–13, 119, 126
DES *see* Data Encryption Standard
detachment 41
deterrence 30
Deterrence Theory 19
Dewey, John 9, 11, 39–40, 44–53, 81,
    132–3
dialogic democracy 42–3
dichotomy of public vs private 81, 92,
    135–6
diffusion of security 4–6
digital cryptography 58
digital encryption 1–4, 58–79; access to
    3–4; basic principles 60–63;
digital era 82–3, 85–6
digital espionage 1
digital self-defence 103
dissidence 139
domestic data 140–41
drones 7, 19
Dropbox 4
drug dealing 71, 74
DuPont, Quinn 65

e-commerce 63–4, 69
eavesdropping 60, 137–8
'eclipse of the public' 49, 51
economic security 86–90, 97–8, 112–13
economy as threat and protection object
    88–92
EFF *see* Electronic Frontier Foundation
Electronic Frontier Foundation 4, 29,
    59–60, 69, 90, 121, 139
elliptic-curve cryptography 64
encryption 80–110; *see also* digital
    encryption; national security and
    encryption
'encryption as threat' narrative 97–8
end-to-end encryption 4, 94–5
engaging with multiple publics 45–7, 52
Enigma 63
ENISA *see* European Union Agency for
    Network and Information Security
Escalante, John J. 94
escrow systems 68–9, 72–3, 75
Escrowed Encryption Standard 68
espionage 84, 98–9
European border security 7
European Union Agency for Network
    and Information Security 124–5

experimentation 49
exporting encryption 12, 67–8, 75, 136–7
expression of will 50
extra-legal surveillance 99

Facebook 5, 27, 90
Fachgruppe Krypto des Gesellschaft für Informatik 29, 102, 118
FBI *see* Federal Bureau of Investigation
Federal Bureau of Investigation 11–12, 28–9, 68–72, 74–5, 80, 85, 88, 95, 100, 111, 113, 120–22, 126
fetishisation 26, 52
fight for privacy 100
FISA *see* Foreign Intelligence Surveillance Act 1978
Five Eyes intelligence pact 71
Foreign Intelligence Surveillance Act 1978 66–7, 140–41
Foucault, Michel 97
framing effects 44, 52
Frankfurte Allgemeine Zeitung 28
Fraser, Nancy 9, 53
fraud avoidance 64
function of security 97

Gaycken, Sandro 28, 85, 115
genetically modified food 42–3
Gheciu, Alexandra 9
Glaser, Barney 25
global encryption 61, 123
global interconnectedness 50–51
'going dark' 80, 93–7, 121
Google 4, 27, 72, 87, 90–92
Grounded Theory 11, 18, 25, 28, 133
Gürses, Seda 112–13, 126

hacking 3, 6, 11, 28–9, 58–60, 64–5, 69–71, 75, 99, 103, 118–19, 123, 133, 137
hacktivism 58
Harding, Sandra 41
Härting, Niko 28, 115
Hayden, Michael 70
Hess, Amy 80, 96
historical context of encryption 58–79
Hobbes, Thomas 138
human dignity 118
humans and technology 114–18
100% fail-safe system 114–15
hybrid forums 39–45, 51, 53
hyper-technologisation 112

illegal mass surveillance 97
illegitimate state action 99
implementation of digital encryption 3–4
incalculable risk 85–6
individualisation of security 105, 113
industrial espionage 3
infrastructure breakdown 23
intelligence gathering 66–7, 71
international data 140–41
internet 20, 23, 27, 58–69, 73, 80–82, 98–9, 101
interoperability 64
interpreting texts 26–31; method of analysis 29–31; selection of material 27–9
invasion of privacy 87–8, 96–7, 102–4, 140–41
iPhone 12, 70, 100, 120–21
issue formation 45–7, 53–4

Johnson, Jeh 89

Kaspersky, Eugene 84, 115
Kreye, Andrian 118
Kurschildgen, Pascal 28, 100

lack of democracy 50–51
Lascoumes, Pierre 39
Latour, Bruno 26
Laurent, Brice 44
Lavabit 73
law enforcement 70–71, 86, 93–4, 125–6, 137
law-and-order principles 154
laws of nature 114
leaks *see* Snowden, Edward
Leander, Anna 25
legitimacy 9, 25, 40–41, 59, 91, 126, 140–41
letter scrambling 61
Leviathan 138
Levison, Ladar 73
liberal democracy 50–51
liberty 137–8
Lippmann, Walter 45
Locke, John 138
Lyon, David 102

maintaining anonymity 73
malicious activity 74
Mandau, Markus 85
marginalisation 43–4
Marres, Noortje 11, 39, 45–9, 52–3, 132–4

Marx, Gary T. 114–19
mass media analysis 28, 31–2, 80–81,
  100–101, 113
mass surveillance 7, 22, 80–81, 85, 91–2,
  95–6, 112
material participation 39, 45–7, 52
material politics 7, 18, 21–4, 46–7
May, Timothy C. 1, 58–9
meaning of encryption 59–60
'mess' of research 25
metaphors 25–6, 45, 74, 95, 97
method of analysis 29–31
methodological individualism 48
methodology of studying material
  politics 24–31; interpreting texts
  26–31; tenets 24–6
methods of knowledge production 121–2
Microsoft 90–92
modality of participation 46
Moore, D. 74
moral character of cryptography 117

narrative of state in control 97–8
National Health Service 74
National Institute of Standards and
  Technology 61, 124–6
National Research Council 68
National Security Agency 11, 64, 70–71,
  74–5, 91, 95, 98–100, 116, 123
national security and encryption 80–119;
  conclusion 104–6; double role of
  encryption 86–104; risks in cyberspace
  81–6
naturalisation 51
nefarious activity 73–4
Neocleous, Mark 138
neoliberalism 104
neopositivsm 24–5
networks 23–4, 48–50, 63–4, 82–4, 102;
  see also internet
Netzpiloten 29, 90
Neumann, Linus 28, 99
neutral citizen 44, 140
new forms of publicness 53–4
new materialism 19
New York Times 28, 999
Newman, Janet 52, 136
NHS *see* National Health Service
9/11 66, 92–3
NIST *see* National Institute of Standards
  and Technology
Niva, Steve 8
non-deliberative theory of democracy 46
normative questions 45

NRC *see* National Research Council
NSA *see* National Security Agency

Obama, Barack 29, 82–3, 118–19
Öffentlichkeit 48
offline life 58
one-click encryption 101
ontological trouble 46
open coding 29–30
opposition to institutions 49
Orwellian future 67
out-of-control state 98–104

paedophilia 94–5
panopticon 97
perfect forward secrecy 117
pervasive technology 58–9
PGP *see* Pretty Good Privacy
PKI *see* public-key infrastructure
plaintext 61
political battlefield of encryption 66–74
political situations 8–9, 22–3, 31–2
politicisation *see* depoliticisation
politics of encryption 60, 74–5
politics of publicness 132–43; publicness
  as concept 134–41
politics of technology 6–8, 18–37
pollution 42
popular culture 28
possibility of resistance 100
possible future worlds 41–2
post-ANT framework 48–9
post-quantum encryption 125
pragmatism 18, 39, 45–6, 51
pre-social individual 48
Pretty Good Privacy 67–8
principles of encryption 60–63
privacy 4–6, 66–74, 100, 104–6
problem of issue formation 45–7
procedural deficit 43
propaganda 53–4
Public and Its Problems, The 47
public: concept of 8–10; Dewey's theory
  of democracy 47–52; hybrid forums,
  technological controversies 40–45; in
  technological society 38–57; problem
  of issue formation 45–7; publicness
  52–3;understanding of science' 40; vs
  private 81, 92, 135–6
public-key cryptography 62, 64–5
public-key infrastructure 62–3
publicity stunts 72, 135; *see also* Apple
publicness 8–10, 52–3, 104–6, 136–7; and
  privacy 104–6; and state 136–7

public–private partnerships 65
PUS *see* 'public understanding of science'

quantum computing 100, 123, 125

ransomware 74, 89, 102
Rational Choice Theory 48
reconfiguring security 5
reduction of threat impact 85–6
reflexivity 48–9
regimes of citizenship 140
regulating technology 59
reification of ideas 52
'research in the wild' 41
researching material politics 24–31
researching politics of technology 18–37
rethinking democracy 40
'return of the public' 38
RFID chips 5, 21
Rid, Thomas 74
Rider, Karina 105
Ries, Uli 116, 119
risks in cyberspace 29–31, 81–6
Rogaway, Phillip 117
role of technology 19–21
Rouse, Joseph 41
RSA cipher 62
RSA Data Security 67–8

S-boxes 64
Saco, Diana 58
sacred institutions 50
San Bernadino attack 70–71, 120
scepticism 43, 103
Schneier, Bruce 84, 105, 116
Schröder, Thorsen 103
scrutiny of secret service activity 139
secluded research 40–42
secrecy 60, 139
secret services 100, 139
securitisation theory 105
security provision 1–2
security/liberty 137–9
selection of material 27–9
self-driving vehicles 3, 64
self-reflection 49, 51
session keys 117
sexism 43
side-effects of technology 114–18
singing politics into existence 86

Skype 139
Snowden, Edward 2, 4, 27, 71, 73, 83, 87, 90–91, 95, 97, 99–100, 113, 123, 133, 139–40
social networks 102
socio-technological controversies 31–2, 41, 52, 116, 141
sovereignty 134–41
'space of government' 122
spatial metaphors 45
SSL/TLS protocol 61
Staatstrojaner 59, 999
stability 50
standard-setting agencies 6, 134
state and publicness 136–7
state as threat 98–9
statehood 10
state–security nexus 105
Steinschaden, Jacob 90
Strauss, Anselm 25
subversive technology 58
Süddeutsche Zeitung 28
Suite A 123
surveillance 3–6, 66–8, 98–9, 102–4, 139–41
surveillance–industrial complex 4
symmetrical encryption systems 62

techno-fallacies 114, 119
technocratic logic 116
technological context of encryption 58–79
technological controversies 40–45
technological society 8, 23–4, 38–57, 103, 111; publics 38–57
technologisation of security 69
technologised environment 22
telecommunication systems 68
terrorism 8, 22–3, 58, 67, 70–71, 92–3, 95, 120, 141
The Onion Router 73–4
theory of democracy 11, 45–54
theory of technology 113–20
'thick' concept of politics 6, 39, 52
Thorsen, Einar 91
3DES method 61
TOR *see* The Onion Router
TOR Project 70, 73–4
traceability 73
tracking 92
traditional politics 75, 137
traffic analysis 63, 73
translation 41
'true' public 53–4, 140

truth vs non-truth 19–20
Turing, Alan 63

UN *see* United Nations
uncertainty 114–18
understanding encryption controversies
    31–2
United Nations 82
unreadability of encryption 60–61
use of private data 5
user activation 102–4

violation of privacy 66; *see also* invasion
    of privacy
vulnerability of technology 69, 84, 116

Wacquant, Loïc 24–5
Walker, Rob 139
WannaCry 74

war technology 63–6
War on Terror 71
Washington Post 28, 85, 89–91, 93–4,
    101
West, Sarah Myers 92
Westin, Alan 4–5
WhatsApp 4, 92, 139
Wikipedia 123
Wired 28, 103, 121
wiretapping 66–7, 71
Wood, M. 101
world politics 19–21
World War II 63

xenophobia 43

Yahoo 91

Zimmermann, Phil 67–8, 75, 136